Nathaniel Jarvis Wyeth
Oregon Dreamer

Tom Calderwood
Bend, Oregon

In February, 1834, a report by Harvard Botanist Thomas Nuttall was read to a meeting of the Academy of Natural Sciences of Philadelphia. It began:

"This collection was made wholly on the returning route of Mr. W[yeth] from the Falls of the Columbia [Celilo] to the first navigable waters of the Missouri... The number of the species and their interest to the botanist will therefore be duly appreciated, and particularly when it is known that this was the first essay of the kind ever made by Mr. W. and yet I can safely say, that besides their number, there being many duplicates, they are the finest specimens probably, that ever were brought from the distant and perilous regions of the west by any American collector." (A Catalogue of a Collection of Plants, Nuttall 1834).

The plants in question were gathered during Nathaniel Wyeth's 1832-1833 expedition to the Oregon Country, a trip undertaken not for science, but for business. Wyeth was a native of Cambridge, Massachusetts. In 1831 when the siren song of the Oregon Country called, he had a good job in a successful ice harvesting firm. Envisioning a green pasture for shrewd Yankees, the thirty-year-old Wyeth organized a commercial venture that failed economically, but set the stage for the Oregon Trail, which brought a tide of American settlement to the region. Wyeth was not trained as a naturalist but he was a founding member of the Massachusetts Horticultural Society and was already building up an orchard when he departed for the West.

Cambridge lies across the Charles River from Boston. Fresh Pond in Cambridge is one of many kettle ponds left in New England by retreating glaciers. It freezes over in the winter and by the early 1800s the ice was being cut and sold commercially. Nathaniel Wyeth invented tools and techniques that greatly increased the productivity of the ice trade, making a major export business possible. He was, by 1831, making $1200/year and had further business interests that brought him, he said, "as much more."

The prime commercial opportunity in the Pacific Northwest was, of course, the fur trade, but Wyeth did not set his aspirations so narrowly. He planned ventures in fishing and tobacco farming and told his brother, "My plan ... is to raise 50 men to go out to that country so early as to leave St. Louis on the 1st May 1832 for the purpose of following the trade of that country in all its branches for which we deem ourselves competent." The enterprise would be a joint stock company, with the participants as shareholders but himself as sole director. It was envisioned to last five years, after which a further, larger, venture could be initiated. In the trade business, Wyeth would rub elbows with operators of various corporate models. Such players included the Hudson's Bay Company (a multinational corporation), the Rocky Mountain Fur Company (a partnership), the expedition of Benjamin Bonneville (backed by the wealthy John Astor), and "free trappers" (operating on their own). Wyeth was different. As an entrepreneur, he raised capital to run a company of his own. Predictably, his correspondence in preparation for the trip

Moss phlox: Wyeth's specimen at Harvard, left, and Nuttall's illustration, right. Cropped image from herbarium sheet *Phlox muscoides*, GH00091720. Gray Herbarium of Harvard University. The illustration was originally published in the Journal of the Academy of Natural Sciences, volume VII part I, 1834; the digital version was acquired from the University of Toronto Gerstein Science Information Centre (https://www.biodiversitylibrary.org/page/24676677#page/79/mode/1up).

involved much juggling of money, of which there never seemed to be enough. Wyeth raised all the cash for his expedition himself; the other participants earned their shares by virtue of their labor.

Wyeth's Yankee practicality as an ice industrialist stands in sharp contrast to his blind optimism about prospects in the Oregon Country. He knew little about the territory itself, or the difficulty of the journey to reach it. He knew equally little about beaver trapping, salmon fishing, and tobacco farming. Wyeth simply assumed that opportunities abounded and that the fruits were there for the taking.

Wyeth's first expedition

Wyeth's first expedition left Independence, Missouri on May 12, 1832, traveling in company with experienced fur traders. Early in the journey west, men began deserting. In the Rocky Mountains, there was a major firefight with the Blackfoot tribe. As the party pressed on, men continued to quit and by November Wyeth was a company of one, staying at Fort Vancouver in the hospitality of John McLoughlin. Wyeth had reasoned that if men were engaged as shareholders, they would be willing to work hard for an eventual reward. But he failed to anticipate that if his venture ran into difficulties (dimming the likelihood of an eventual reward), men who were not being paid wages would see no reason to stay with the operation. Wyeth collected but few beaver pelts, and there would be no fishing for salmon and no cultivation of tobacco. Furthermore, the ship that Wyeth sent to bring trade goods from Boston to the West Coast wrecked on a Pacific island. Failure was complete, and he wrote in his journal, "I am now afloat on the great sea of life without stay or support but in good hands i.e. myself and providence and a few of the H.B. Co. who are perfect gentlemen."

Wyeth did not commit many botanical observations to his journal, but he did make two notes regarding trees in the Pacific Northwest. On October 7, 1832, on the way to Fort Walla Walla, he wrote, "...we passed through an immense forest of pine of different kinds and unknown to us altho[ugh] very similar to some of ours..." These would likely have been ponderosa, lodgepole, and Western white pines, the latter being much like the Eastern white pine that graces New England. On November 29, 1832, venturing up the Willamette River, he writes, "On the bottoms there is consid[e]rable oak of a kind not found in the states but of excellent quality for ship building and is the only kind of oak found in the country of the Columbia." This was Oregon white oak (*Quercus garryana*).

Wyeth managed to induce two of his former party to join him on the arduous return journey—one not to be taken alone. It was on this return trip that Wyeth collected the specimens that were celebrated by Thomas Nuttall, who was curator of the botanical gardens at Harvard. Some of Wyeth's specimens may be viewed online at https://kiki.huh.harvard.edu/databases/specimen_index.html (search for collector "N. J. Wyeth").

Thomas Nuttall named two flowering plants in Wyeth's honor: *Clematis wyethii* and *Wyethia helianthoides* (white mule's ears). He noted that the former appeared closely related to *Clematis douglasii*. Of the latter, he wrote: "In the valleys of the Rocky Mountains, near Flat-Head river, in rich plains. Flowering

Wyeth buckwheat: Wyeth's specimen at Harvard University, left, and Nuttall's illustration, right. Cropped image from herbarium sheet of *Eriogonum heracleoides*, GH00036370. Gray Herbarium of Harvard University. The illustration was originally published in the Journal of the Academy of Natural Sciences, volume VII part I, 1834; the digital version was acquired from the University of Toronto Gerstein Science Information Centre (https://www.biodiversitylibrary.org/page/24676677#page/79/mode/1up).

Wyethia helianthoides grows abundantly in the Ochoco Mountains; hybrids between it and *W. amplexicaulis* have creamy flowers (*Wyethia* x *cusickii* Piper). Photo by Cindy Roché.

pushed towards home immediately. The indicated parcel of plants collected from the west coast, however, was apparently not with him. A letter written the same day to Leonard—to whom the package was supposed to go—makes no mention of it. And Wyeth wrote Nuttall, "I have sent" rather than "I am sending" the package. Such a parcel, dispatched from the Fort Vancouver area, could have gone overland through Canada (courtesy of the Hudson's Bay Company) or by ship from the lower Columbia. Either way, it was apparently lost.

Wyeth's second expedition

By November 1833, Wyeth was back in Cambridge, making preparations for a *second* trek to Oregon, his enthusiasm for the fur business undiminished by the dismal results of the first trip. He had learned many lessons from his expedition. He started as a complete tenderfoot and returned as a man reasonably experienced in the ways of the West. He also observed a key business fact that propelled him towards a second venture: the American fur companies were procuring their trade goods in Missouri and carrying them overland to the Rocky Mountains. They paid high prices for the goods and incurred steep costs in transport. Wyeth realized that there was another way. Goods could be bought cheaper on the East Coast and sent by ship to the Columbia. Once there, horses to transport the goods inland could be had at much lower prices than in Missouri and the eastward route to the interior was shorter and less dangerous than the trek westward. Furthermore, the ship could then be used to carry furs back to Boston and the cost of the voyage met by bringing back cured salmon to complete the load in the ship.

about the beginning of June. The root of this plant is, I believe, no less than those of the preceding genus [Espeletia helianthoides], employed for food by the aborigines after fermentation [in the ground] and exposure to a parboiling heat."

Wyeth kept a letter-book, and on July 4, 1833, he penned the following to Thomas Nuttall:

> *"Dear Sir: I have sent through my brother Leon[ar]d of N. York a package of plants collected in the interior and on the western coast of America somewhere about Latt. 46 deg. I am afraid they will be of little use to you. The rain has been so constant where I have been gathering them that they have lost their colors in some cases, and they will be liable to further accident on their route home."*

This letter suggests that Wyeth sent two packages of plant specimens and the first one was lost. His letter indicates that some plants were collected "on the western coast," but when Nuttall published the new species (Nuttall 1834) all of the collections were from east of Celilo. And Nuttall made no mention of any specimens damaged by water. In July 1833, when Wyeth wrote this letter to Nuttall, he was in the Rocky Mountains (present day Idaho), on his way back to Massachusetts. His return was briefly delayed by an unexpected opportunity to start a joint trapping operation with Benjamin Bonneville. Because this venture would have occupied Wyeth for another year, his letter would have been carried east by someone else. The arrangement with Bonneville quickly fell apart, after which Wyeth

Wyeth's insights were perfectly valid, but in his ardor for big profits in the West he seemed to forget the obstacles that had plagued his first expedition. A key factor in any successful business venture is the proper estimation—and mitigation—of risk. He would again be faced with unreliable employees, uncertain communications, loss of pack animals (exhaustion, straying, injuries), Indian attacks, accidents, illness, and maritime casualties. And from the start of his second foray, he had less money than he needed, even in his own opinion, and spent it faster than he anticipated.

The shining hope that motivated the 1834 expedition was a contract he had signed with members of the Rocky Mountain Fur Company at the rendezvous of fur traders in 1833. He agreed to bring them a supply of trade goods—goods that he would still need to carry overland,

White mule's ears (*Wyethia helianthoides*) illustration by Thomas Nuttall. The illustration was originally published in the Journal of the Academy of Natural Sciences, volume VII part I, 1834; the digital version was acquired from the University of Toronto Gerstein Science Information Centre (https://www.biodiversitylibrary.org/page/24676677#page/79/mode/1up).

but that he could purchase in Boston and New York at favorable prices. This would give Wyeth's new firm, the Columbia River Fishing and Trading Company, an immediate source of profit. A ship with a cargo of Wyeth's own trade goods would be dispatched to meet him on the Columbia and fulfill his larger plans for harvesting fur and salmon.

Wyeth departed Independence, Missouri for the Oregon Country on April 28, 1834. He did not serve as the botanist on this trip, for Thomas Nuttall was now a member of the party. The Harvard curator had resigned his post to take this extended journey. He and John Townsend rode out ahead to collect specimens before the plants were crushed under the hooves of the numerous horses and cattle of the large expedition. After an arduous trek, Wyeth

reached the Green River in Wyoming on June 19th. There, he met the Rocky Mountain Fur Company, which reneged on the contract, refusing to pay for the trade articles. In his journal, Wyeth complained of the "scoundrels" he had to deal with, but he should have known that fur traders were high-risk customers. He pushed further west and used the unsold articles to set up his own trading post, Fort Hall, on the Snake River, not far from modern-day Pocatello, Idaho. This location eventually became an important stop for wagon trains headed to the Oregon Territory.

Wyeth arrived at Fort Vancouver on September 14, 1834 and learned that his ship, the *Mary Dacre*, was anchored further downstream. The crew were to have set up a salmon fishery and begin curing fish from the summer run. But it was not to be. The vessel, while sailing in the Atlantic off the coast of South America, was struck by lightning. Captain James Lambert wrote:

> *...at 6 o'clock in the morning, lightning struck our main top-gallant mast head, stove that into splinters, descended down the top-mast splitting that into pieces, through the main cap splitting the main mast head and down the mast, between the eyes of the lower rigging and mast to the deck, where it exploded with a most tremendous report. It ripped up the deck*

NATHANIEL J. WYETH.

Nathaniel J. Wyeth. Portrait courtesy of the Museum of the Mountain Man, Pinedale, Wyoming.

amidships, the partners and coaming off the mast and went into the hold, where the electrick fluid separated again, one part flying aft through the Cabin door, staving everything in pieces in its course, but the other most fatal ball passed out through the Brig's side close to the water's edge, starting off all the bends amidships, and a great quantity of the copper, leaving the Brig on fire inside, and filled with sulphuric smoke.

It's ironic that one of Wyeth's reasons for favoring ocean delivery of trade goods to the Columbia was that subsequent overland transport from Vancouver to the Rockies was safer than carriage there from Missouri. He ignored the hazards of the sea, which was quite strange given that the vessel for his first expedition was a total loss. The *Mary Dacre* limped all the way to Valpariso, Chile, where the crew largely deserted while slow and expensive repairs were made to the ship. The ship reached the Columbia River only shortly before Wyeth did, and far too late for the salmon run that, by chance, proved miserably small anyway. With no fish to subsidize the transport of beaver skins, there was no point in sending the ship back to Boston that year and Wyeth resolved to keep the vessel at hand in the hope that the following year would bring a piscine cargo.

In addition to botanist Nuttall, Wyeth's party included ornithologist John Townsend who kept a journal covering more than just avian matters. After the expedition reached Fort Vancouver, Townsend joined an excursion up the Willamette River, where he noted, "...the timber is generally pine [fir?] and post oak, and the river is margined in many places with a species of willow with large oblanceolate leaves like those of the peach, and white on their under surface [*Salix scouleriana*]." He further wrote that the islands in the river have oaks, but no pines. Townsend's journal is generally much more literary than Wyeth's, and he captured some of the emotions of the exploration:

> *"None but a naturalist can appreciate a naturalist's feelings—his delight amounting to ecstasy—when a specimen such as he has never before seen, meets his eye, and the sorrow and grief which he feels when he is compelled to tear himself from a spot abundant with all that he has anxiously and unremittingly sought for."*

Wyeth proceeded to set up his own permanent installation on the Columbia. He chose the site of an abandoned native village on Wapato (now Sauvie) Island. This would serve as the base for the next year's salmon fishing, and he set his men about the tasks of building barrels and storehouses. Fort William resembled a country village with carpenters, smiths, and other artisans as well as numerous domestic animals. Wyeth traveled up the Willamette River to find land suitable for farming. While the purpose of the farm was to supply produce for Fort William, Wyeth curiously did not consider Wapato itself as a location (it still hosts farms today). Wyeth settled on a parcel near

what was known as French Prairie, not far from modern-day Champoeg, roughly 40 miles from his home base on Wapato Island.

With affairs in the lower Columbia left to his lieutenants, Wyeth turned his attention back to the interior country and beaver. Carrying the trade articles brought by the *Mary Dacre*, he led a group of his men back upriver to Fort Walla Walla. But once he got there, twelve of the members deserted, taking considerable property with them. Wyeth spent weeks chasing after them, a trek that brought him back down the Columbia with a major side excursion up the Deschutes River in the dead of winter. Storm-bound one night with his trapping companion, he wrote in his journal:

> *"It makes two individuals feel their insignificance in the creation to be seated under a blanket with a fire in front and 3½ feet of snow about them and no telling when it will stop. Tonight 'tis calm and nearly full moon. It seems to shine with as much indifference as the storms blow and whether for weal or woe we two poor wretches seem to be little considered in the matter. The thoughts that have run through my brain while I have been lying here in the snow would fill a volume and of such matter as was never put into one, my infancy, my youth, and its friends and faults, my manhood's troubled stream, its vagaries, its aloes mixed with the gall of bitterness and its results: viz under a blanket hundreds perhaps thousands of miles from a friend, the Blast howling about and smothered in snow, poor, in debt, doing nothing to get out of it, despised for a visionary, nearly naked, but there is one good thing plenty to eat: health and heart."*

He finally caught up with some of the scofflaws at Fort Vancouver, recovering none of his goods, but catching some beaver along the way. Back at Fort William, Wyeth turned to preparations for fishing. He had told his backers that a large haul could be caught, "...when the proper mode is found." But Wyeth underestimated the effort

Fresh Pond in Cambridge, looking west-northwest. Fort Vancouver lies three thousand miles in the distance. Photo by the author.

Wyeth's legacy

The species named in honor of Nathaniel Wyeth by Thomas Nuttall, British botanist at Harvard, were collected on Wyeth's 1832 Expedition's return journey (eastbound). Wyeth's connection to Nuttall was his cousin, John Wyeth, who was a neighbor of Thomas Nuttall in Cambridge. Nuttall published the names for the 1833 collections in the Journal of the Academy of Natural Sciences Philadelphia in 1834: *Clematis wyethii, Eriogonum heracleoides, Wyethia helianthoides*, and *Espeletia amplexicaulis*.

Nuttall had arrived in Philadelphia in 1807 and taught for a time at Harvard. He spent 33 years exploring the North American continent during which time he was a member of the Academy of Natural Sciences in Philadelphia. Nuttall did not return east with Wyeth on his second expedition, instead he sailed from the west coast to the Hawaiian Islands in December 1834. He returned to the Pacific Northwest in the spring of 1835 and continued his botanical explorations for another year before sailing back to Boston from San Diego in May 1836. By sheer chance he booked passage on a ship where one of his former student was serving as sailor. (Richard Henry Dana, Jr. wrote *Two Years Before the Mast* based on this voyage). Many species in Oregon bear Nuttall's name, but his botanical discoveries are necessarily another story.

Wyeth's collections whetted the appetite of naturalists, and when the second expedition left Independence in May 1834, in addition to Nuttall, it included ornithologist John Kirk Townsend. The entourage also included 70 men, 250 horses, and Baptist missionaries Jason and Daniel Lee with their cattle, so the botanists

Large populations of *Wyethia amplexicaulis* grow on seasonally wet clay soils in the Ochoco Mountains. Photo by Jennifer Curtis.

found it necessary to ride far out ahead to obtain samples before they were trampled. The route they followed became known as the Oregon Trail; missionaries were soon followed by settlers. More than 270,000 emigrants eventually stopped at Fort Hall on their journey. Although Wyeth did not cross the Mississippi again, he encouraged many American settlers to go west, speaking with a voice of authority as someone who had been there.

Plants named for Wyeth

Even though Montana can claim most of Wyeth's type specimens, a number of those species are also part of Oregon's flora. Given the fluidity of taxonomic names, it is not surprising that many of the species that Thomas Nuttall named for Nathaniel Wyeth no longer bear his name.

The type specimen for Wyeth clematis (*Clematis wyethii* Nutt.) was collected "towards Flat-Head river, and in flower on the 25th of June 1833." In 1885 it was considered a subspecies of Douglas clematis (*Clematis douglasii* subsp. *wyethii* (Nutt. ex Torr. & A. Gray) Kuntze). The currently accepted name for this clematis is *Clematis hirsutissima* Pursh., sugarbowls. In Oregon, this clematis grows primarily in the Blue Mountains, extending southwest into the Ochoco Mountains.

The buckwheat that Wyeth collected at the "sources of the Missouri" in early June 1833, *Eriogonum heracleoides* Nutt., still bears the common name Wyeth buckwheat.

Wyeth collected white mule's ears (*Wyethia helianthoides* Nutt.) about the beginning of June 1833 "in the valleys of the Rocky Mountains, near Flat-Head River, in rich plains." In Oregon, white mule's ears grow in the Blue Mountains, extending into the mountains of the northern Basin and Range in southeastern Oregon.

Wyeth collected northern mule's ears (*Wyethia amplexicaulis* (Nutt.) Nutt.) in 1833 in "habitat about Flat-Head river." Nuttall's original name was *Espeletia amplexicaulis* in 1834. This species is widely distributed in eastern Oregon.

Narrow-leaved mule's ears (*Wyethia angustifolia* (DC.) Nutt.) was originally published as *Alarconia angustifolia* DC. by Augustin de Candolle using a collection by David Douglas as the type specimen. Nuttall moved it to *Wyethia* in 1840. Narrow-leaved mule's ears grows in western Oregon.

—Cindy Roché and Kareen Sturgeon.

needed to catch a whole ship's worth of fish, especially by his inexperienced men. In the end, he landed less than half the salmon expected.

Wyeth shipped out what beaver and fish he had aboard the *Mary Dacre* and returned overland to Massachusetts. Along the way, he still grasped for ways to turn his business around, to no avail. He eventually sold Fort Hall and the goods at Fort William to the Hudson's Bay Company, but the value of these assets was insufficient to cover the $20,000 he invested in the two expeditions.

Return to his roots

Back in Cambridge for good, Wyeth returned to ice harvesting at Fresh Pond, prospering through the success of his own company. That he should have made his fortune close to home in a familiar trade, rather than far afield in unproven ventures, seems wholly appropriate for a Yankee businessman. The New England work ethic, after all, did not call for flights of fancy. But for other explorers, the western territory would still yield great returns.

John Townsend wrote in his *Narrative of a Journey* (Townsend 1999 OSU reprint) of the potential for scientific discovery:

> *"What valuable and highly interesting accessions to science might not be made by a party, composed exclusively of naturalists, on a journey through this rich and unexplored region! The botanist, the geologist, the mammologist, the ornithologist, and the entomologist, would find a rich and almost inexhaustible field for the prosecution of their inquiries, and the result of such an expedition would be to add most materially to our knowledge of the wealth and resources of our country, to furnish us with new and important facts relative to its structure, organization, and natural productions, and to complete the fine native collections in our already extensive museums."*

References

Hardee J. 2013. *Obstinate Hope: The Western Expeditions of Nathaniel J. Wyeth, Vol. 1.* Sublette County Historical Society, Pinedale, Wyoming.

Hardee J. 2018. *Hope Maintains Her Throne: The Western Expeditions of Nathaniel J. Wyeth, Vol. 2.* Sublette County Historical Society, Pinedale, Wyoming.

Nuttall T. 1834. *A Catalogue of a Collection of Plants made chiefly in the Valleys of the Rocky Mountains or Northern Andes, toward the sources of the Columbia River, by Nathaniel B.[sic] Wyeth.* Journal of the Academy of Natural Sciences of Philadelphia, Vol. VII part I.

Townsend J. 1999. *Narrative of a Journey Across the Rocky Mountains to the Columbia River.* Reprint, Oregon State University Press, Corvallis, Oregon. 290 pp.

Oregon Historical Society. 1899. *The Correspondence and Journals of Captain Nathaniel J. Wyeth 1831-1836.* University of Oregon Press, Eugene, Oregon. 262 pp. (https://openlibrary.org/books/OL23291298M/Correspondence_and_journals_1831-6)

Tom Calderwood is a native of Portland, Oregon. Educated at Benson High School and the Massachusetts Institute of Technology, he pursued a career in software engineering in Greater Boston. Now retired, he lives in Bend where he writes and practices astronomy. His work has appeared in Sky and Telescope Magazine and the Thoreau Society Bulletin. He is a member of the High Desert Chapter of NPSO.

Marah Mysteries: Confusion over Wild Cucumber

Frank Callahan
Central Point, Oregon

sidebar on **Cucurbit Pollinators and Insect Visitors** by
Lincoln R. Best
Oregon Bee Atlas, Oregon State University
Corvallis, Oregon

How many species of "wild cucumber" grow in Oregon? This question has intrigued me for decades. I've lived and botanized in southwestern Oregon since 1970, when I bought 20 acres in the foothills southeast of Gold Hill in Jackson County (35 miles north of California on Interstate 5). Although I noticed considerable variation in leaf morphology in the *Marah* populations I encountered, I didn't give it much thought because the Oregon Flora included only one species in the gourd family (Cucurbitaceae): Oregon bigroot (*Marah oregana*).

Marah oregana leaves and male flowers. Photo by Frank Callahan.

The bigger picture

I've long been fascinated by members of the gourd family (Cucurbitaceae). While serving in Vietnam, I encountered bitter melon (*Momordica charantia*) and Vietnamese luffa or bath sponge (*Luffa aegyptiaca*), whose immature fruits are often cooked and eaten. Gourds are a large and diverse family, comprising some 98 genera and approximately 975 species. Humans are most familiar with the species that are edible vegetables/fruits, including the Old World cucumbers (*Cucumis*) and melons (*Citrullus*) and New World summer and winter squash (*Cucurbita*). The Cucurbit species in Oregon are all toxic and extremely bitter, thus without food value for humans. In fact, in 1855 California botanist Albert Kellogg chose the name Marah from the Bible, Exodus 15:22-25, which describes Moses finding a spring after three days without water in the wilderness, but the water was so bitter the people could not drink it. They named the spring marah, the Hebrew word for bitter. The bitterness is caused by the presence of terpenoid compounds called cucurbitacins, which are found in all parts of the plant. Cucurbitacin B functions as a defense against herbivores.

The mystery of the *Marah* species

In the past few years, I started collecting specimens for herbarium vouchers and found that the leaf variations were coupled with even larger differences in the fruits, flowers, and seeds. Thus, I set out to solve what I considered the mystery of the *Marah* species in southwestern Oregon. In short order my collections revealed that there are three species of *Marah* in the wooded hills near my property. Two of these, California man-root (*M. fabacea*) and Taw man-root (*M. watsonii*), were previously undocumented in Oregon and constitute range extensions into southwestern Oregon from the California Floristic Province. I suspect that the recent discovery of these populations can be explained because of the limited access to their locations. Most of the area is privately owned by armed individuals who vigorously deny access not only to their own property but also to public land that can be reached only by crossing their land. In some ways, this denial is more limiting to botanists than when David Douglas was collecting. He had only to deal with inclement weather, grizzly bears, poison oak, rugged terrain, fire, and the difficulty of preserving and transporting specimens.

In addition to my discoveries of new *Marah* species in Oregon, I also found balsam apple (*Echinocystis lobata*), which was not in the Oregon Flora. I found it in 2020 while surveying for *Asclepias incarnata* along the Snake River at the boundary between Idaho and Oregon. Although previously undetected due to its inaccessibility, this population of balsam apple is not unexpected because it is adjacent to large native populations in Idaho.

In addition to Oregon bigroot recorded in Peck (1961), four other cucurbits currently appear on the

The root of *Marah fabacea,* which I excavated from a landslide, cleaned, and placed on its side, measured 58 cm long. Photos by Frank Callahan.

Closeup of the surface structure of the root of *Marah fabacea.* Note the embedded stone upper right corner of image.

website (Oregonflora.org), but without floristic treatment. Two *Marah* species (*M. fabacea* and *M. watsonii*) are shown as "exotic?" and two other taxa (West Indian gherkin (*Cucumis anguria*) and balsam apple (*Echinocystis lobata*)) as "not naturalized." *Cucumis anguria* was collected as a garden escape in 1918 on a vacant lot in Portland and did not persist. Both *Echinocystis* and *Marah* species are commonly called "wild cucumber." As they are both in the cucumber family, this shared common name leads to confusion.

Differences between *Marah* and *Echinocystis*

The genera *Marah* and *Echinocystis* differ primarily in their method of seed germination and their root systems. *Marah* has a tuberous root system and hypogeal[1] seed germination. In contrast, *Echinocystis* has a fibrous root system and epigeal seed germination. Since *Echinocystis* commonly grows in riparian zones with ground moisture available throughout the year, the fibrous root system serves it well. *Marah* is commonly known as bigroot or man-root because its thickened tubers grow to enormous size and can be as heavy and large as a man, weighing as much as 100 kg. *Marah*'s massive tuberous roots store water for the dry season, a useful adaptation to Mediterranean climates with long, hot, droughty summers. The tubers contain saponins and have been used as a soap or ground up and placed in water to stupefy fish.

It is not easy to view the roots of *Marah* species. Because the tubers often extend more than five feet deep in the ground, either a backhoe or excavator is needed to

---

[1]In hypogeal germination, the cotyledons remain below the soil surface during germination. In epigeal germination, the cotyledons emerge above the soil surface.

Outline of typical leaf shapes of *Marah* and *Echinocystis*. Illustration by Cindy Roché.

unearth them. Occasionally near the coast, the roots are exposed by high tides or landslides. In the very wet year of 1965, I found plants in the Coast Range of Coos County where a landslide exposed the tubers that "floated to the surface" of the moving soil.

The above-ground parts of *Echinocystis lobata* resemble *Marah* species, with lobed leaves, vining stems and coiled tendrils. Since both are often called wild cucumber, casual observers often assume they are the same. However, a quick comparison of leaf morphology sets them apart: *Echinocystis* leaves have five lobes, each tapering to a single acute point, so that the leaf resembles a 5-pointed star. *Marah* leaves have lobes with multiple points, giving a more rounded appearance.

Three *Marah* species in Oregon

The three species of *Marah* in Oregon are long-lived perennial vines that trail on the ground or climb by simple or

Side view of female flower and developing fruit of *Marah oregana*. Photo by Cindy Roché and Robert Korfhage.

2 mm

Top view of female flower of *Marah oregana*. Photo by Cindy Roché and Robert Korfhage.

2 mm

Top view of male flower of *Marah oregana*. Photo by Cindy Roché and Robert Korfhage.

branched tendrils that cling to upright vegetation. The above-ground vines die back each year to the large, deep-seated tuberous roots. Plants survive for a long time, sometimes living over 100 years. Leaves are simple with shallow lobes in Oregon bigroot and California man-root and deeply lobed in taw man-root. Plants are monoecious with male and female flowers produced on separate stalks from the same leaf axils. Staminate flowers are borne in racemes or narrow panicles, while pistillate flowers are mostly solitary. The five white petals are fused at the base, with the calyx reduced to inconspicuous lobes inserted at the sinuses of the petal lobes. The corolla is densely covered with short glandular hairs. In male flowers, the corolla is pale-green, and the stamens are fused with the anthers and twisted together in a mass to form an orange center. The female flower has an inferior ovary, from which the fruit, a capsule, develops below the flower. *Marah* produces a single crop of fruits each year in summer. Capsules generally bear prickles and are usually four-valved, splitting at the tip, where the seeds fall out. In comparison to the rigid spines on the 17 cm capsules of *Marah horridus* in the central Sierra Nevada foothills of Fresno County, prickles are soft in the species in Oregon. The dried capsules are attractive and often used in dried flower arrangements (but handle carefully!).

In the garden, members of Cucurbitaceae are notorious for hybridizing. Do wild species of *Marah* also hybridize? I have seen no indication of hybrids in the wild, even though Oregon bigroot and California man-root grow together and all three *Marah* species in southwestern Oregon grow within the foraging distances of potential pollinators, which makes cross-pollination possible. Stocking (1955) found no evidence of hybridization between species of *Marah* in nature or in experimental gardens.

Oregon bigroot (*Marah oregana*)

Oregon bigroot leaves are suborbicular, up to 13 cm across, with rounded to cordate bases and three to nine shallow lobes. Tubers can grow to enormous size, up to 100 kg. The corolla is cup-shaped. Seeds form in ellipsoid capsules that are up to 7.5 cm long. Capsules are densely covered with soft prickles and usually beaked at the tip. The dark brown seeds are disk shaped, 16 mm in diameter and 8 mm thick.

Of the three species found in Oregon, Oregon bigroot has the widest distribution, extending from southwestern British Columbia to the Central Valley and San Francisco Bay area in California. At the northern limits of its range in British Columbia it is known as coast manroot and is classified as endangered (COSEWIC 2009). By growing so far north, *Marah oregana* is exceptional in Cucurbitaceae, which is primarily a family of the tropics and subtropics. Success in the northern latitudes is attributed to the development of the deeply buried enormous tubers (Stocking 1955).

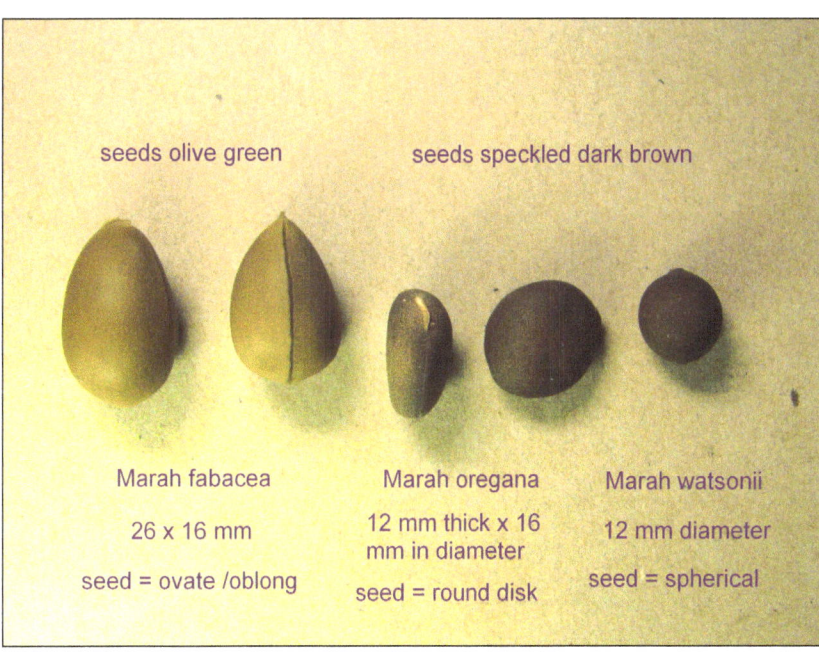

seeds olive green   seeds speckled dark brown

Marah fabacea

26 x 16 mm

seed = ovate /oblong

Marah oregana

12 mm thick x 16 mm in diameter

seed = round disk

Marah watsonii

12 mm diameter

seed = spherical

Beaked fruit of *Marah oregana*. Photo by Frank Callahan.

Seeds of the three *Marah* species in Oregon, showing differences in size, shape, and color. Photo by Frank Callahan.

In Oregon, this species extends eastward to Umatilla County, with isolated populations in Baker County. Although some sources indicate it as native to Idaho, the reports from Bingham County, Idaho, are in error and are actually collections of *Echinocystis lobata* (Boise State U.35513 and NY Botanical Garden Barcode 01154036).

Partly because there were no specimens of *Marah oregana* from Klamath County in Oregon herbaria, in July 2016 two colleagues (Steve Northway, Diana Wageman) and I set out to relocate Newberry's Oregon bigroot above the shoreline of Klamath Lake. Our exploration included Eagle Ridge, a peninsula in the middle of Upper Klamath Lake. We followed the narrow, single lane road all the way to its end at the northern tip of the peninsula. We found *Marah oregana* trailing in the grasses in very dry habitat in several places along this road. At the end of the road, we descended a cattle trail to the shoreline and found *Marah* plants vining among bitter cherry (*Prunus emarginata*) and Klamath plum (*P. subcordata*) that grew near a small concrete tub with flowing warm water.

Oregon bigroot's favorite habitat is Oregon white oak (*Quercus garryana*) woodland or savanna, but it is also common in the transition zone forest, where conifers mix with broadleaf trees. In southwestern Oregon, near Blackwell Hill and Gold Hill, *Marah oregana* thrives on well-drained, sandy or silt loam soils derived from granite and gabbro parent rocks. *Marah oregana* does not perform well on ultramafic soils. Other than the serpentine form of *Marah watsonii*, *Marah* species are generally not soil specific. Oregon bigroot is also adapted to a variety of other habitats, as I've found it growing on the exposed barren rocky summit of Soda Mountain in Jackson County (1856 m elevation) and near sea level at Harris Beach State Park near Brookings. It is found in riparian sites inland along the Columbia River drainage and in coastal habitats,

Marah oregana in typical open woodland habitat with *Quercus garryana* and *Cercocarpus betuloides*. Photo by Frank Callahan.

such as windblown coastal shrublands, grassy headlands, prairies and forest openings (including Douglas-fir plantations). In coastal shrublands dominated by salal (*Gaultheria shallon*), evergreen huckleberry (*Vaccinium ovatum*), and crowberry (*Empetrum nigrum*), it vines up through the shrubs and spreads out over the canopy. It is found in coastal forests with Sitka spruce (*Picea sitchensis*), Douglas-fir (*Pseudotsuga menziesii*) and bigleaf maple (*Acer macrophyllum*) and on the headland and seacliffs, where its large leaves overtop the grasses. Henderson commented on his collection from the Winchuck River in 1929 that it was "very common everywhere."

Leaves and male flowers of *Marah fabacea*. Photo by Frank Callahan.

California man-root (*Marah fabacea*)

In 1859, French naturalist Charles Victor Naudin (1815-1899) described a plant grown in the botanical gardens in Paris from a large root from the San Francisco Bay area in California (Naudin 1859). He named it *Echinocystis fabacea* Naudin. Usually, the type specimen is collected from a wild population which becomes the type locality, but in this case the location of the wild population is not precisely known so the type specimen is the garden plant without a type locality.

California man-root leaves resemble those of Oregon bigroot in shape and number of lobes, but leaves of *M. fabacea* are about 2.5 cm smaller. Tuber size is also similar to that of Oregon bigroot. California man-root's corolla is rotate, not cup-shaped as in the other two species. The densely prickly capsule is up to 4 cm long, globose to slightly depressed. The name *fabacea* refers to the seed, which looks more like a seed in the pea family (Fabaceae) than one of Cucurbitaceae, which tend to be flattened. The olive-green seed is oval in both width and length, about 16 mm thick by 22 mm long, much larger than other *Marah* species.

California man-root is the most common man-root in that state. Its range extends from San Diego County north through Siskiyou County. Until I reported this species in Jackson County in the vicinity of Gold Hill it was considered a California endemic. In Oregon, this species grows at about 425 m elevation in Oregon white oak woodlands but isn't found in madrone (*Arbutus menziesii*) woodlands. It appears to be limited to silt loam soils derived from decomposed metasedimentary rock; it has not been found on clay soils. In California, it is especially abundant in the chaparral zone and also inhabits rock and scree zones in the mountains (very rarely over 1500 m elevation) and open fields in the Sacramento Valley.

Taw man-root (*Marah watsonii*)

The type specimen, originally named *Echinocystis watsonii* Cogn., was collected in the vicinity of Placerville, California. It was deposited at the California Academy of Sciences, but did not survive the San Francisco fire of 1906 (Stocking 1955). Until I found it in southwestern Oregon, taw man-root was described as a California endemic, occurring in the foothills around the north end of the Central Valley. The populations closest to Oregon are in the foothills north of Redding, a distance of about 160 miles from Gold Hill.

The most common habitats in California are rocky canyons in the lower foothills to transition zone woodlands where conifers mix with broadleaf trees; it seldom grows in open fields. Taw man-root also grows on ultramafic or serpentine influenced soils where it appears to be a distinct ecotype because the plants differ markedly in appearance from those growing on other soils. For example, on the ultramafic soils at Walker Ridge in Lake County, California, leaves are skeletal, with a glaucous, waxy lower surface. This growth form is often found in serpentine environments because it protects leaves from intense light and heat reflected from the exposed soil surfaces typical of such environments. In contrast, plants growing on non-serpentine soils in the shade of other vegetation exhibit softer, lush green leaves with more surface area.

Fruit of *Marah fabacea* opened up to expose the four seeds. Photo by Frank Callahan.

In Oregon, taw man-root grows in and around stands of California buckeye (*Aesculus californica*), yet this association with buckeye is not found in California. In Oregon California buckeye might provide shade that lowers ground temperatures and retains soil moisture. Vining into California buckeye also might afford taw man-root some protection from herbivory as the toxic leaves of buckeye are avoided by browsing animals. Normally the cucubitacins protect *Marah* foliage from grazers. However, during drought years in California the need for forage overwhelms all food preferences and cattle completely consume *Marah* plants in overgrazed fields.

Taw man-root leaves are broader than long, strongly lobed with deep sinuses to over half the width of the leaf. Overall leaf size varies with the fertility of the site. On deep fertile soil, leaves are usually 10 cm wide. Leaves on plants growing on shallow, rocky soils may be half this size. Unlike other *Marah* species, vines of taw man-root are reported to be nearly hairless, with a glaucous gray-green color (Calscape https://calscape.org/Marah-watsonii-(Taw-Man-root)?srchcr=sc57f656fa1e59c). However, this observation may be from serpentine ecotypes. Taw man-root tubers are much smaller than the other two *Marah* species, in the range of 4.5 to 7 kg. Like Oregon bigroot, the corolla is cup-shaped, not rotate as in California man-root. Capsules are about 4 cm long, round but not depressed; the surface can be smooth or sparsely spiny. The dark brown seeds are like tiny marbles, 13 mm in diameter.

Balsam apple or wild cucumber (*Echinocystis lobata*)

Echinocystis is a monotypic genus, with the name derived from the Greek *echinos* (hedgehog) and *kystis* (bladder), referring to the spiny fruits. The type locality of *Echinocystis lobata* is in western Pennsylvania, along the Ohio River. Many references indicate that it is native across much of North America, excluding California, Nevada and a block of southeastern states (USDA PLANTS, Vascular Plants of the Pacific Northwest). It is common on the Idaho side of the Snake River and also grows in Washington and British Columbia (Hitchcock *et al.* 1977). There is some historical inconsistency on whether it is native in the Pacific Northwest. For example, the *Illustrated Flora of the Pacific States* describes it as occurring in "Thickets and waste places, eastern United States and Canada westward to Montana but evidently escaping cultivation and appearing sporadically in the Willamette Valley, Oregon, and eastern Washington and western Idaho" (Abrams and Ferris 1940).

The first known collection of *Echinocystis lobata* in Oregon was by M.W. Gorman in Portland on August 30, 1905 (WS104066). Two early collections made in the riparian zone of the Link River near Klamath Falls include one by M.K. Small on August 18, 1936 (SOC 23721) and one by E.R. McLeod on September 3, 1952 (SOC 16020). In 2015 Eileen Laramore collected *Echinocystis*

lobata along the Umatilla River at Gate Camp on the Oxbow Property near Hermiston (Umatilla County). In 2020, when Tom Fealy and I surveyed for *Asclepias incarnata* along the Snake River in Malheur County, we found *Echinocystis lobata* growing along a 48-km stretch of the Snake River at the Oregon/Idaho border, from the Fort Boise Wildlife Area to north of Ontario. It isn't surprising that this population in Malheur County was not discovered earlier, as all of the land here is privately owned and the only access to the riverbanks is by boat, launching at the sportsman's access points on the Snake River. In these floodplain habitats, it vines up through indigo bush (*Amorpha fruticosa*) and, as a very aggressive climber, overwhelms the tops of the bushes. We collected voucher specimens for the herbaria at Oregon State University (OSC) and Southern Oregon University (SOC).

Unlike *Marah* species, which are perennials with tuberous root systems, *Echinocystis* is an annual vine with fibrous roots, re-establishing from seed every year. Plants

Typical leaf of *Marah watsonii* on serpentine-influenced soil. Photo by Frank Callahan.

Typical leaf of *Marah watsonii* on normal soil. Photo by Frank Callahan.

trail across the ground or climb other vegetation using curling tendrils originating in the leaf axils. Leaves are usually 10 to 15 cm long and wide with five triangular lobes. Like *Marah*, *Echinocystis* is monoecious, with male and female flowers borne on separate stalks in the same leaf axils. The fragrant, white to pale yellow-green flowers have six narrowly lanceolate petals about one cm long that are fused at the base. The petals are densely covered with straight hairs. Male flowers are borne in upright panicles and have three stamens, in which the filaments and anthers are more or less connate. Female flowers are borne singly on a stalk at the base of the male panicles. They have an inferior ovary that develops into a sub-globose capsule, 2.5 to 5 cm long, with weak spines. Seeds are flat and ovate, 16 mm long. *Echinocystis* produces fruits from spring until early winter, making it easy to collect seeds. As in *Marah*, the seeds require cold stratification prior to germination. Seeds are readily consumed by rodents, which may aid in their dispersal.

Fruit of *Marah watsonii* showing curved spines. Photo by Frank Callahan.

Relationships with insects for native cucurbits and garden cultivars

Marah is the native host plant for the Western spotted cucumber beetle or 12-spot beetle (*Diabrotica undecimpunctata*). This native beetle eats *Marah* leaves down to the veins, which can end growth for the season. Fortunately, *Marah* is not permanently damaged by this because it has enormous reserves in those underground tubers. These reserves allow it to endure periods when beetle populations are high and to produce seed again when beetle populations are low. Human cultivation of non-native cucurbits complicates the relationship between beetles and native plants. Gardeners blame native *Marah* for beetles in their cucurbits, but an abundance of cucumbers, squashes, and melons in gardens contributes to beetle damage in native wild cucumber populations.

Another insect, the striped cucumber beetle (*Acalymma vittatum*), feeds on both native and cultivated cucurbits. As they go from plant to plant, these beetles transmit a bacterial wilt (*Erwinia tracheiphila*). Infection occurs when adult beetles defecate on leaves where they have been feeding and bacteria enter the plant through the damaged areas. *Echinocystis* is susceptible to bacterial wilt. When cultivated and wild cucumbers grow in proximity to each other, it doesn't bode well for either one when it comes to diseases.

Never Stop Discovering!

Solving the mystery of odd-looking wild cucumbers in Oregon revealed three species that had not been included in the *Flora of Oregon*. I'm sure that this is not the last remaining discovery of unreported species in Oregon, so if you see something that looks unusual or doesn't key properly in the *Flora*, pursue it. You might be making a new discovery.

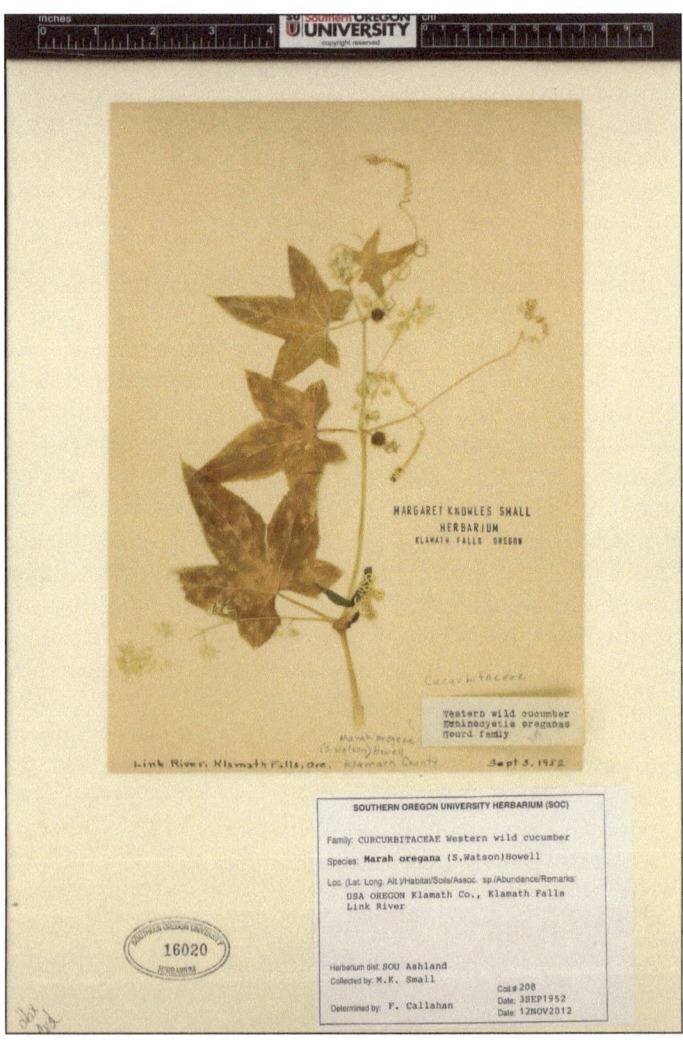

Herbarium sheet of *Echinocystis lobata* collected by Margaret Knowles Small in Klamath County in 1952, misidentified for decades as *Marah oregana*.

Cucurbit Pollinators and Insect Visitors

Lincoln R. Best

Bumble bees and host-specific pollinators called squash bees are the most frequent pollinators of large-flowered North American native species in the genus *Cucurbita*. Both also pollinate cultivars of *Cucurbita pepo* (pumpkin) and *Cucurbita foetidissima* (buffalo gourd), the bumble bees as generalists and the squash bees by adaptation to new hosts. Volunteers with the Oregon Bee Atlas (OBA), a project to inventory all the bees statewide, found the first squash bee ever recorded in Oregon, *Peponapis pruinosa*, which was collected from squash flowers in a community garden in Ashland (Best *et al.* 2019). After this discovery in 2018, the OBA directed surveyors to look for squash bees only in large-flowered squashes and pumpkins, because cucumbers are too small to accommodate the squash bees (OSU Extension 2020). This squash bee coevolved with large-flowered native cucurbits in the American Southwest so the two life cycles are perfectly aligned: bee feeding habits (collecting pollen and nectar) are adapted to the morphology and phenology (timing of flowering) of cucurbits. Early in the morning, squash flowers open and offer prodigious quantities of sugary nectar to attract the bees. In the male flowers, the bee is passively dusted with pollen, which it transfers to the next female flower that it visits. Squashes, pumpkins, and gourds are the sole pollen hosts for the females. During the first few hours after sunrise male bees dart between flowers searching for mates; by noon they are fast asleep in the withered flowers.

Since cucumber flowers are too small for *Peponapsis pruinosa*, flowers of native *Marah* and *Echinocystis* would also be too small. Pollination ecology of native cucurbit species has not been studied in Oregon, but information on bees that visit *Marah oregana* flowers has been documented by the OBA, which is part of the Oregon Bee Project. Volunteers collected 28 bee specimens from *Marah oregana* in the Willamette Valley in 2018 and 2019 (Best *et al.* 2021, 2022). Bee visitors to *Marah* included seven genera of native bees, mostly small carpenter bees in the genus *Ceratina*. In 2022, bee visitors included the small carpenter bees (*Ceratina*), mining bees (*Andrena*), mason bees (*Osmia* and *Protosmia*), a small sweat bee (*Lasioglossum*), cuckoo bee (*Nomada*), long-horned bee (*Eucera*) and a honeybee (*Apis*).

These visitation records do not necessarily indicate that these insects pollinate *Marah oregana*. Anecdotal reports mention that *Marah oregana* flower nectar attracts small bees in southwestern Oregon (https://www.amateuranthecologist.com/2015/05/floral-visitors.html) and that *Marah fabacea* in California is a nectar source for butterflies in The Presidio (National Park) (https://www.parksconservancy.org/park-e-ventures-article/one-cool-cucumber-marah-and-its-frightening-fruit) and for the green hairstreak butterfly (*Callophrys rubi*) in San Mateo County Parks (https://friendsofedgewood.org/california-manroot). *Ceratina* bees have also been reported as the most frequent visitors to *Marah* flowers in California (Schlising 1966). Small black ants, honeybees, and small beetles (Stocking 1955) and bumble bees, *Andrena* bees and gnats (Moldenke 1976) have also been recorded as flower visitors to *Marah* plants in California. The white flowers of *Marah* species produce nectar and remain open at night (Callahan, pers. comm.), so moths are also possible pollinators. I have observed significant numbers of owlet (Noctuid) moths associated with *Marah* vegetation; they may also visit the flowers for nectar. Andrew Moldenke, Research Professor at Oregon State University, has observed clearwing moths (Sessiids) visiting both sexes of flowers (Moldenke, pers. comm.).

Leafcutter bee (*Megachile* sp.). Photo by Lincoln R. Best, Oregon Bee Atlas.

Black-tailed bumble bee (*Bombus melanopygus*). Photo by Lincoln R. Best, Oregon Bee Atlas.

The naming game for *Echinocystis* and *Marah*

The earliest collection of Oregon bigroot in the Oregon Territory was in 1834 by Scottish botanists John Scouler, David Douglas, and William Fraser Tolmie, "On the Oregon [Columbia River] from near its mouth to Kettle Falls." British botanist Joseph Dalton Hooker (1817-1911) named this collection *Sicyos angulatus* Hook. (Fl. Bor. Am. 1:220, 1834). In 1840 John Torrey and Asa Gray (who had been Torrey's student) changed the name of Scouler's collection to *Sicyos oreganus* Torr. & A. Gray.

Confusion arose when Torrey used the name *Megarrhiza oregana* for this species in 1855 without a type specimen and later referred to the same plants as *Echinocystis*. The 1855 Pacific Railroad Survey's physician and botanist, Dr. John Strong Newberry[1] reported it "on the shores of Klamath Lake and banks of the Willamette River, Oregon Territory, in August & September in fruit" (Newberry 1857). He used the name *Megarrhiza oregana* (literally, Oregon bigroot). The first time the name *Megarrhiza oregana* appears in print is in the catalog of plants of the Expedition of the Pacific Railroad Reports, but "no character is given, or any synonym or citation of a description" (Greene 1887). Most of the plants from the 1855 expedition were sent to American botanist John Torrey (1796-1873), the leading authority on the flora of North America at the time. Torrey's name is appended as the authority for the name *Megarrhiza oregana*, indicating "that he was entertaining the thought of founding a genus upon these plants; an opinion which it is evident that he shortly afterwards relinquished" since a few years later in his report for the Wilkes Expedition[2] (1838-1842), Torrey referred to the same plants as *Echinocystis* (Greene 1887).

Meanwhile, events in California added to the confusion. American physician Albert Kellogg (1813-1887), the first resident botanist in California, collected specimens in 1854 near San Francisco that he named *Marah muricatus* Kellogg, based on gigantic fleshy roots and

[1]See article in *Kalmiopsis* 22:10-19.
[2]See article in *Kalmiopsis* 12:16-24.

other differences (Kellogg 1854). Kellogg commented in his description, "The significance of the name we have chosen would be better understood by perusing Exodus XV: 22-25." His genus name is significant because his type was later determined to be *Marah oreganus*. To add yet more confusion, Sereno Watson published *A Revision of Megarrhiza* in the 1875 Proceedings of the American Academy which, rather than being a revision of an established genus, was the first characterization of it (Watson 1875). Neither Newberry's first publication of the name nor the revision had a type specimen or description.

Thus, the two type specimens for Oregon bigroot are the 1834 collection from the Columbia River somewhere between the Pacific Ocean and Kettle Falls, Washington, and the 1854 collection near San Francisco (US National Herbarium 1906).

Oregon bigroot becomes *Echinocystis* in 1878 and returns to *Marah* in 1898

In 1878 Belgian botanist Célestin Alfred Cogniaux (1841-1916) made *Marah* one of the three sections of the genus *Echinocystis*, thus it became *Echinocystis oregana* Cogn. (Mem. Cour. Ac. Belg. 28:87, 1878). Using Constantin Rafinesque's 1808 name *Micrampelis*, American Botanist Edward Lee Greene (1890) placed the *Marah* species in that genus: *Micrampelis oregana* (Torr. & Gr.) Greene, a name which apparently did not gain acceptance but lingers on as a synonym.

In his 1898 Flora of Northwest America, John Thomas Howell renamed *Echinocystis oregana* (Torr. ex S. Watson) Cogn to *Marah oreganus* (Torr. ex A. Gray) Howell. The Hebrew word for bitter came from an older Latin root, *Amarus,* also meaning bitter. In Hebrew, Mar is the masculine form and Mara is the feminine form. *Marah* species described as *Echinocystis* originally had feminine epithets but because *Marah* is feminine, the specific epithets have now been corrected to *Marah oregana* and *Marah fabacea*.

References

Abrams L. Ferris R.S. 1940. *Illustrated Flora of the Pacific States* Vol. 4. Stanford University Press, Stanford California. pp. 66-67.

Best L, Engler J, Feuerborn C, Larsen J, Lindh B, Marshall CJ, Melathopoulos A, Kincaid S, Robinson SVJ. 2022. Oregon Bee Atlas: Wild bee findings from 2019. Catalog of the Oregon State Arthropod Collection. Preprint. DOI: http://dx.doi.org/10.5399/osu/cat_osac.6.1.4906.

Best L, Feuerborn C, Holt J, Kincaid S, Marshal CJ, Melathopolous A, Robinson SVJ. 2021. Oregon Bee Atlas: native bee findings from 2018. Catalog of the Oregon State Arthropod Collection. 5 (1) 1-12. DOI: https://doi.org/10.5399/osu/cat_osac.5.1.4647.

Best LR, Marshall CJ, Red-Laird S. 2019. Confirmed presence of the squash bee, *Peponapis pruinosa* (Say, 1837) in the state of Oregon and specimen-based observational records of *Peponapis* (Say, 1837) (Hymenoptera: Anthophila) in the Oregon State Arthropod Collection. Catalog: Oregon State Arthropod Collection. 3(3) p 2–6DOI: http://dx.doi.org/10.5399/osu/cat_osac.3.3.4614

COSEWIC. 2009. COSEWIC assessment and status report on the Coast Manroot *Marah oreganus* in Canada. Committee on the Status of Endangered Wildlife in Canada. Ottawa. vi + 28 pp. (www.sararegistry.gc.ca/status/status_e.cfm).

Greene EL. 1890. A Series of Botanical Papers. *Echinocystis & Megarrhiza*. Pittonia 2:129.

Hitchcock CL, Cronquist A, Ownbey M, Thompson JW. 1977. *Vascular Plants of the Pacific Northwest Volume 4*. University of Washington Press, Seattle, WA. pp. 481-482, 486.

Kellogg A. 1854. *Marah muricatus*. Proceedings of the California Academy of Sciences. San Francisco. 1:38–39.

Moldenke, A. R. 1976. California pollination ecology and vegetation types. Phytologia 34:305-361.

Naudin CV. 1859. *Marah fabacea* Naudin. Ann. Sci. Nat., Bot. Sér. 4, 12: 154-156.

Newberry JS. 1857. Botany Report. *In* Explorations for a Railroad Route, from the Sacramento Valley to the Columbia River, made by Lieutenant R.S. Williamson, Corps of Topographical Engineers, assisted by Lieutenant H.L. Abbott, Corps of Topographical Engineers: Reports of Explorations and Surveys, to Ascertain the Most Practicable and Economical Route for a Railroad from the Mississippi River to the Pacific Ocean. US Government Printing Office, Washington, DC. Part 3 p. 74.

Peck ME. 1961. *A Manual of the Higher Plants of Oregon*. 2nd ed. Binsfords & Mort.

OSU Extension 2020. The Great Oregon Squash Bee Hunt. (https://extension.oregonstate.edu/gardening/pollinators/great-oregon-squash-bee-hunt)

Schlising RA. 1966. Reproductive ecology of plants in the genus *Marah* (Cucurbitaceae). PhD Dissert., Univ. Calif., Berkeley, California.

Stocking KM. 1955. Some Taxonomic and Ecological Considerations of the Genus *Marah* (Cucurbitaceae). Madroño 13(4):113-137.

Watson S. 1875. Botanical Contributions VI. On the Flora of Guadalupe Island, Lower California. Proc. American Academy of Arts and Sciences. pp. 105-147. (https://www.jstor.org/stable/pdf/20021459.pdf)

Frank Callahan is a veteran author for Kalmiopsis having published six previous articles. His botanical experience spans six decades: he collected his first herbarium specimen near the Metolius River in 1963 where he was working as an information host with his grandparents at the Allingham Guard Station. Since then, he has botanized in all of the western United States west of the Rocky Mountains and on the east coast of Florida. He did 16 expeditions into Mexico. Farther afield, he has botanized in Viet Nam (while serving in the USMC), in Japan, and in Spain, Croatia, and Italy. His interest in the largest specimens of trees led to discovering over 80 National Champions and about 28 Oregon State Champions. In 2021 was awarded the Maynard C. Drawson Memorial Award for Heritage Tree Stewardship. He taught the conifer classes at Siskiyou Field Institute in Selma for five years. He is a long-time volunteer in the Southern Oregon University Herbarium in Ashland and has led numerous field trips for the Siskiyou Chapter. He considers his home in the foothills of the Klamath-Siskiyou Mountains, with their complex geology that supports many rare and unusual plants, a botanical paradise.

Lincoln Best was born in the Cariboo Region of central British Columbia, and raised in southern Ontario, Canada. He earned a Bachelor of Science with Honours in Zoology from the University of Guelph and studied the taxonomy of bees and wasps around the world. Lincoln serves as the lead Taxonomist for the Master Melittologist Program and Oregon Bee Atlas (OBA) at Oregon State University. The OBA is a community-science initiative of trained Master Melittologists who inventory and monitor the biodiversity of native bees in the state with a focus on documenting bee visitation of flowering plants.

Ricegrasses: Rocks, Rodents, and Rarity

Cindy Roché
Bend, Oregon

James David
Soil Scientist, Ochoco National Forest
Prineville, Oregon

Jack Maze
Professor Emeritus, Department of Botany
University of British Columbia, Vancouver, British Columbia

Eastern Oregon is home to two rare endemic ricegrasses. Henderson's ricegrass (*Eriocoma hendersonii*) is a regional endemic, limited to Oregon and Washington, and Wallowa ricegrass (*Eriocoma wallowaensis*) grows only in Oregon. Both taxa are Forest Service "Sensitive Species" and BLM "Special Status Species." Details on other designations of rarity may be found at Natureserve (https://www.natureserve.org/) or Oregon Biodiversity Information Center (https://inr.oregonstate.edu/orbic).

For many years, Henderson's ricegrass was known from only two disjunct populations, one in south-central Washington and the other in the Ochoco Mountains (Hitchcock 1971). Gradually, additional populations were discovered (Vrilakas 1990), extending the range northward in both states, but a large gap remained between populations in Washington and those in Oregon. Then, in the 1970s, Henderson's ricegrass was discovered in Wasco County, Oregon (Winward and Youtie 1976); this placed a dot on the map directly within the wide gap between the two population centers at the time. The known distribution of Wallowa ricegrass has not changed since it was described in 1996. Most of the populations grow in north end of the Wallowa Mountains, with a small outlier group in the Ochoco Mountains 300 km to the southwest.

No one knows why such long distances occur between the two population centers of each species, or why they are rare. Both species grow in scablands communities, commonly on ridges, where the soil is shallow over bedrock. There is no shortage of scabland habitat in the areas

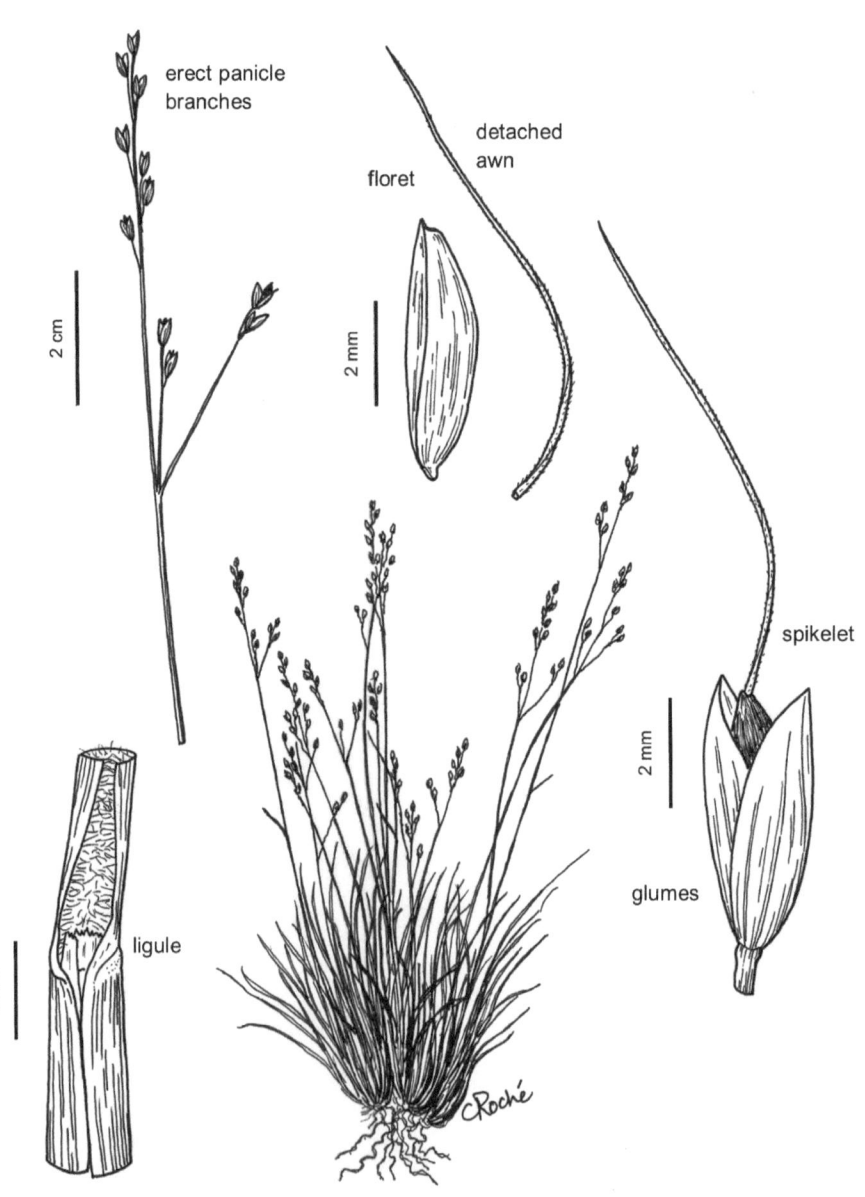

Henderson's ricegrass has erect panicles. Illustration by Cindy Roché.

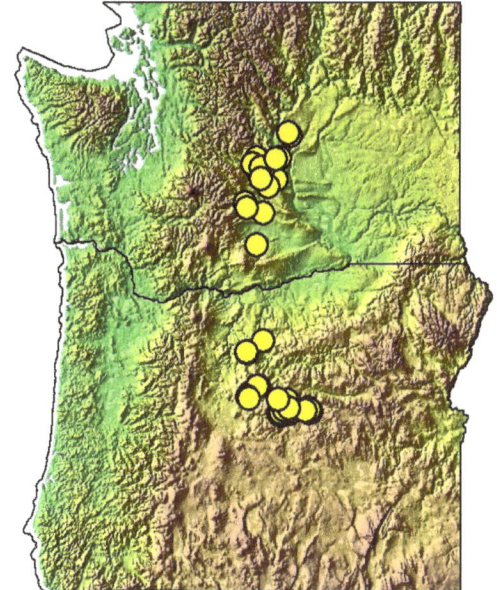

Distribution of Henderson's ricegrass in Oregon and Washington. Map by Katie Mitchell with the Oregon Flora.

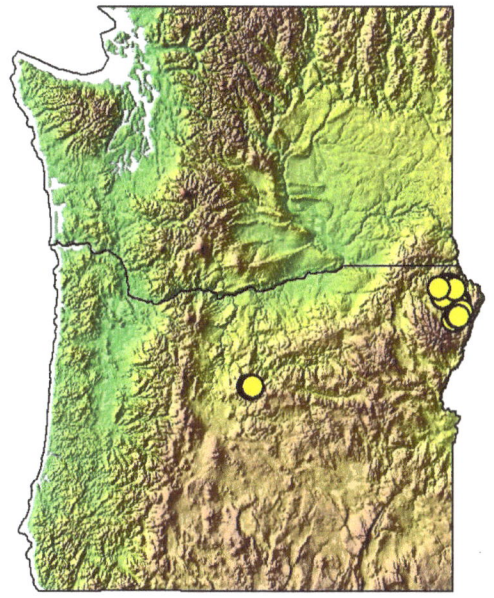

Distribution of Wallowa ricegrass in Oregon. Map by Katie Mitchell with the Oregon Flora.

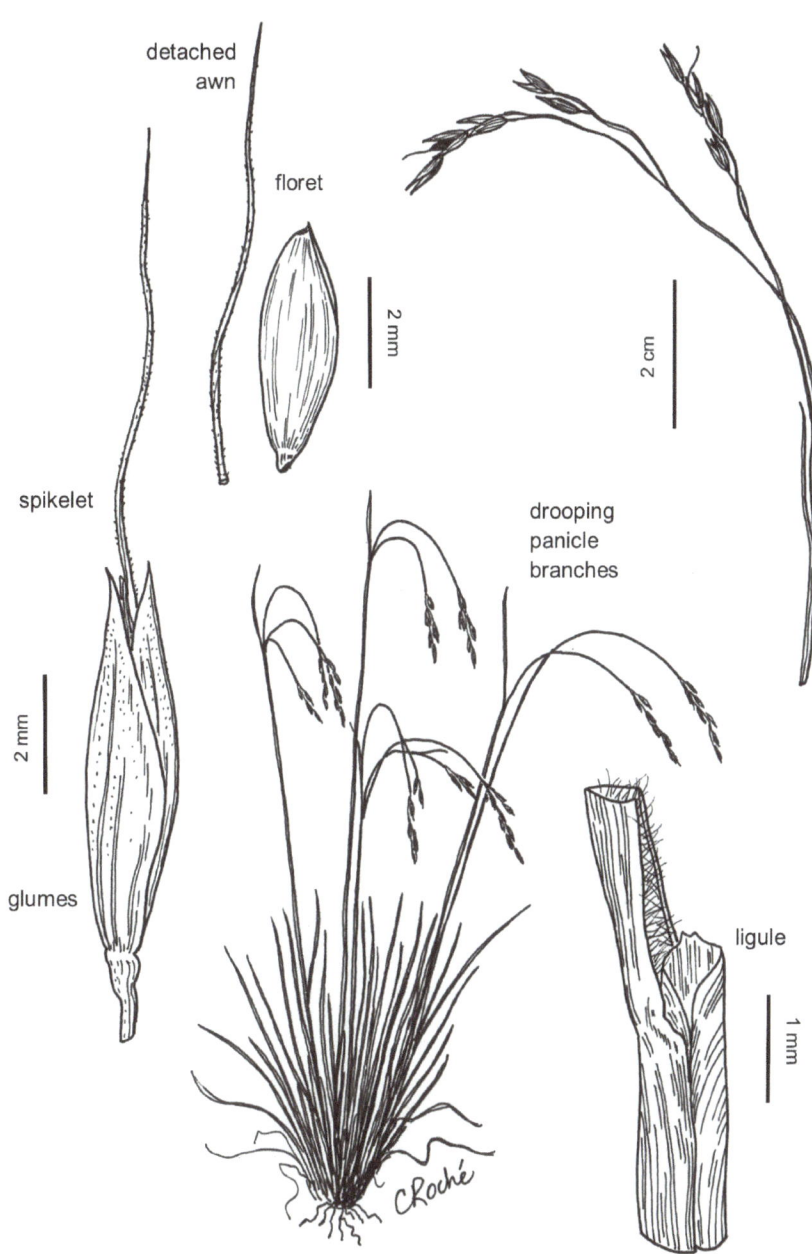

detached awn

floret

2 mm

2 cm

spikelet

drooping panicle branches

2 mm

glumes

ligule

1 mm

CRoché

Wallowa ricegrass has drooping panicles. Illustration by Cindy Roché.

between populations. Scablands are a common landscape feature east of the Cascade Mountains in Oregon and Washington, occupying over 600,000 acres in eastern Oregon alone (Kari Littrell, Natural Resources Conservation Service, pers. comm.).

Binney (1997) recommended an intensive exploration of both biotic and abiotic factors to unravel why Henderson's ricegrass grows where it does. In this article we present our detailed observations of the habitat of Henderson's ricegrass in Oregon and suggest which factors could explain why Henderson's ricegrass is rare. Habitat details are important because it can improve the efficiency of surveys for its presence and provide a foundation for management guidelines to favor survival of the species. We include comparisons with Wallowa ricegrass, its closest relative, because current descriptions of the habitats of the two species are the same: both species are restricted to scablands (non-forested, shallow, rocky soils).

Meet the rare ricegrasses

Henderson's and Wallowa ricegrasses are small bunchgrasses with long, fibrous roots with relatively few branches (see the photos comparing roots of Henderson's ricegrass with those of Sandberg bluegrass on page 23). The stiff, erect leaves are rolled or folded, with fine hairs on the

Name changes in the needlegrass tribe

Ricegrass is a common name for members of the needlegrass tribe (Stipeae) that have short, rounded or oval florets. The species with long, narrow, sharp-pointed lemmas are called needlegrasses. Years ago, most of the needlegrasses were in the genus *Stipa* and the ricegrasses were *Oryzopsis* (which reminds us of the rice-shaped pasta called orzo). In North America, *Stipa* was subsequently divided into *Achnatherum, Hesperostipa, Jarava* or *Pappostipa, Nassella,* and *Piptatherum* or *Piptatheropsis* (Barkworth *et al.* 2007). More recently, most *Achnatherum* species became *Eriocoma* (Peterson *et al.* 2019). For those who detest new names for plants, *Eriocoma* is not a new name. It dates back to 1818 for *Eriocoma cuspidata*, Thomas Nuttall's early name for silk-grass, which is our Indian ricegrass! The name erio (wool) and coma (head of hair) was used for the copious silky hairs on the lemma (Nuttall 1818). The name combination of *Eriocoma hymenoides* was published by Per Axel Rydberg (1912).

Spikelet of silk-grass *(a.k.a.* Indian ricegrass) showing the dense silky hairs on the floret. Photo by Cindy Roché.

Five species in the Pacific Northwest bear the common name ricegrass: Indian ricegrass (*Eriocoma hymenoides*), roughleaf ricegrass (*Oryzopsis asperifolia*), little ricegrass (*Piptatheropsis exigua*), Henderson's ricegrass (*Eriocoma hendersonii*) and Wallowa ricegrass (*Eriocoma wallowaensis*). Henderson's, Indian, little, and roughleaf ricegrass were all at one time *Oryzopsis*. Wallowa ricegrass wasn't recognized until the early 1990s, but keep reading, because we'll share that story shortly. Roughleaf ricegrass doesn't grow in Oregon and little ricegrass isn't particularly rare. Indian ricegrass is common and widespread, especially on sandy sites east of the Cascade Range. This article focuses on Henderson's and Wallowa ricegrasses, the first of which was named 100 years before the second one was recognized.

inner surface of the blades. The basal sheaths become flat and stiff with age and persist to form a coarse shield around the shoots above the crown. The membranous ligules are 0.5 to 1.5 mm long. The inflorescence is a panicle, with spikelets bearing one floret each. The elliptic glumes with pointed tips are longer than the body of the floret. The florets are elliptic and laterally flattened, a shiny, dark brown when mature. A single, curved or slightly bent awn, attached slightly off-center at the tip of the lemma, falls off at maturity. The most obvious difference between the two species is the inflorescence. Henderson's ricegrass has an erect panicle with ascending branches, with the spikelets oriented upward. The panicle of Wallowa ricegrass is drooping with thin, lax branches with the spikelets oriented downward.

Habitat differences between Henderson's and Wallowa ricegrasses

Even though Henderson's and Wallowa ricegrasses occupy what appears to be the same type of shallow soil habitat, they do not grow in the same locations. The maps suggest that their ranges overlap in the Ochoco Mountains, but the closest populations are separated by about 20 km (Farris 2013). Specific characteristics of the Wallowa ricegrass sites have not been studied, but the underlying geology clearly differs at the sites where the two species grow in the Ochoco Mountains. Wallowa ricegrass grows over the John Day Formation (a mixed-up jumble of basalts, tuffs, and rhyolites) dating back 30 to 18 million years. In contrast, Henderson's ricegrass grows over younger formations: mainly Picture Gorge Basalts, but also occurs on some rhyolitic tuffs (Paulson 1977). The Picture Gorge Basalts are 17 individual layers of flood basalts over 400 m (1300 feet) thick. The flows occurred from 16.4 to 15 million years ago in the early to middle Miocene and contain layers of buried ancient soils that developed in intervals between flows.

Existing reports do not describe geologic conditions at Wallow ricegrass sites in northeastern Oregon, but all of the sites are on relatively flat ridges and south-facing slopes north of the Wallowa Mountains proper (Paula Brooks, pers. comm.). Thus, these sites were probably not glaciated during the Pleistocene. Like the unglaciated Ochoco Mountains, this area was probably subject to cycles of soil deposition and erosion (Johnson and Clausnitzer 1992). About 12,000 years ago, eruptions of Glacier Peak blanketed the Blue Mountains and Ochocos with ash, with a second hefty addition of sandy volcanic ash about 7,700 years ago when Mt. Mazama erupted (Fryxell 1965). Both wind and water erosion reworked the deposits of ash and loess to create shallow soil on the ridges and deep soil in the drainages.

The soils on the Wallowa ricegrass sites on the Wallowa-Whitman Forest are very sparsely vegetated, and the rocks are red and pebbly, with a mulch of pea-sized rocks on the surface. Potential habitat is easily detected on

Pattern of scablands (shallow soils) and stringers (deeper soils), both created by erosion and deposition during the Pleistocene in the Ochoco Mountains. Deeper soils support trees and appear green in the photo. Image from Google Earth.

aerial photos, but many locations that looked appropriate on photos did not support plants when checked (Paula Brooks, pers. comm.).

Scabland characteristics in the Ochoco Mountains for Henderson's ricegrass

Scablands are a habitat of harsh extremes. Two factors (the underlying bedrock and a thin layer of clay soil above it) prevent free drainage of water from snowmelt or rainfall, resulting in extreme water saturation and alternating freeze/thaw cycles in late winter and early spring. Summer arrives early on exposed ridges and slopes, creating parched, bone-dry conditions early in the growing season. Humans tend to evaluate soils in light of their own uses, for example, agriculture and livestock grazing. Sites that are "more rocks than soil" were judged worthless by European settlers. But Native Americans valued lithosols as important foraging sites for *Lomatium* and other edible roots. The scablands are also critical habitat for Henderson's and Wallowa ricegrasses.

Henderson's ricegrass grows only on very shallow lithosols, but it doesn't grow on all scablands (if it did, it wouldn't be rare, as the Ochoco Forest has over 250,000 acres of scab/stringer landscape). In Oregon, it grows on relatively few and then only on small parts of those scablands. In our examination of the particular sites where Henderson's ricegrass grows, perhaps the most obvious factor is plant competition. Shallow rocky soils support a variety of plant communities, ranging in vegetative cover from sparse to nearly continuous. The ricegrass consistently grows on sites with sparse vegetation. Binney (1997) found that Henderson's ricegrass is not a vigorous competitor with other species (or itself), and Farris (2013) reported that vegetation in all of the Henderson's ricegrass sites was sparse, averaging 18% total plant cover. Our observations in Ochoco scablands are consistent with these studies: we didn't find Henderson's ricegrass in scablands that support a high vegetative cover.

On the Ochoco National Forest, we found that the most vigorous populations of Henderson's ricegrass grow in extremely rocky sites. Rocks appear to an important factor in ricegrass habitat, but not all rocks are created equal. So next we describe the soil at Henderson's ricegrass locations, the parent material (rocks) from which they were derived, and their effect on the habitat.

"More Rocks than Soil"

Rocks provide numerous benefits; they act as heat sinks, moderating temperature extremes and insulating the soil surface; they funnel water into the soil, preventing surface

Henderson's ricegrass growing in a rock matrix. Photo by Cindy Roché.

erosion and evaporation; they protect plants from herbivory, both above and below ground; they reduce competition from other plants, and provide favorable sites for seedling establishment. Indeed, rocks may provide one of the keys to understanding the rarity of the two ricegrasses.

Soil parent material determines fertility, rock characteristics, and water capacity

Although it was known that scabland soils are very rocky in the surface horizon, David (2013) documented the ways in which particular properties of those rocks influence a wide array of factors important to plant distribution. Henderson's ricegrass in the Ochoco Mountains grows primarily on two types of lithosolic habitats: basalt and rhyolitic tuff. Both sites are very shallow (less than 25 cm), with soil depths ranging from 11 to 18 cm on the basalts and from to 10 to 19 cm on the rhyolitic tuffs. Soil textures are similar between the basalt scablands and rhyolitic tuff scablands (Table 1).

Clearly, both sites provide suitable habitat for the ricegrass, so there must be ways in which the two types of soils compensate for differences in the rock parent material.

This section gets technical, so if you aren't into soils and geology, you have our permission to skip over it.

What are the differences between the two rock types, basalt and rhyolite, and how do they affect plant growth? Basalt is high in magnesium and iron and low in silica and feldspar. Silica increases viscosity; lower silica content makes the lava more fluid, creating fine-grained, black rocks when cooled rapidly. In contrast, rhyolite is rich in silica and low in iron and magnesium. It is highly viscous and typically cools slowly in place as dikes or volcanic plugs or is ejected explosively as ash and cinders. (Obsidian is the product of non-explosive quick cooling of rhyolite.) Rhyolitic tuff is the product of explosive volcanic eruptions, in which ash rains down from the sky, solidifying into relatively soft rock layers. Water, ice, lichens, and humic acid subsequently weathered these laminar layers of ashfall tuffs into small flat rocks. Tuff is high in silica and is at least 75% ash; the other 25% may be particles of volcanic glass and small fragments of crystals and/or volcanic rock and lava. Because of the chemical differences, basaltic soils are three times more fertile than the rhyolitic tuff soils and may strongly affect plant vigor (Farris 2013).

Based on this chemistry, basalt and rhyolite rocks also differ in shape and specific gravity. Shape and specific

Table 1. Comparison of study sites on Picture Gorge basalt and rhyolitic tuff scablands. (adapted from David 2013)

Lithology	Average Elevation (range)	Average Soil Depth (range)	Soil Classification	Closest Soil Series Concept
Basalt	4,729 ft. (4,386 to 5,106)	14.8 cm (11 to 18)	Lithic Argixerolls; loamy-skeletal, mixed, superactive, frigid	Canest Series
Rhyolitic Tuffs	4,375 ft. (4,200 to 4,500)	12.5 cm (10 to 19)	Lithic Argixerolls; loamy-skeletal, mixed, superactive, frigid	Tweener Series

gravity of rocks determine the water storage capacity of the soil and accessibility of plants to herbivores. Basalt rocks are spherical or cube-shaped rocks, classified in order of increasing size as gravel, cobbles, and stones (Table 2). On the rhyolitic sites, the rocks are flat instead of round. The cobble-sized rocks are called channers (Table 2).

Basalt rocks are heavier than rhyolite rocks (except for welded tuff): density of the basalt rocks averages 2.9 g/cm^3 compared to 2.4 g/cm^3 for rhyolite channers or 2.2 g/cm^3 for rhyolitic tuff. The heavier basalt is less mobile than the lighter rhyolite, providing a more stable substrate for plant growth and formation of a cryptogamic crust. Another effect of differences in bulk density is porosity, which is

Table 2. Comparison of A Horizon and surface characteristics of the Picture Gorge basalt scablands and the rhyolitic tuff scablands.

A Horizon characteristics	Basalt	Rhyolitic Tuff
Gravels 2-75 mm (≤ 3 inches)	15 %	
Cobbles 45-250 mm (3-10 inches)	57 %	
Stones 250-600 mm (10-24 inches)	2 %	
Channers (flat rocks) 2-150 mm (≤ 6 inches)		63 %
Flagstones (flat rocks) 150-380 mm (6-15 inches)		6 %
Stones (flat rocks) 380-600 mm (15-24 inches)		0.5 %
Gravel mulch (distinct layer of loose surface gravel)		2 cm (0-4 cm range)
Surface texture	Extremely Cobbly (60-90% cobbles) Sandy Loam	Extremely Channery (60-90% channers) Sandy Loam to Coarse Sandy Loam
Cryptogamic crust thickness	0-10 mm	0-1 mm

Henderson's rice grass on a basalt site. Photo by Robert Korfhage.

Henderson's ricegrass on rhyolitic tuff, showing tuff gravel and channers (Refer to Table 2 for rock sizes). Photo by Robert Korfhage.

Rhyolitic tuff gravel forms a mulch over the soil. Photo by Robert Korfhage.

lower for basalts and higher for rhyolitic tuffs. Available water capacity (AWC) correlates positively with porosity. The tuff sites (higher porosity = greater water capacity) store enough water for plants that normally only occur on sites with higher precipitation. For example, on the Snow Mountain District of the Ochoco National Forest, ponderosa pine normally requires 356 mm (14 in.) or more of annual rainfall on basalt substrates, but it can grow on tuff sites with only 305 mm (12 in.) of precipitation. So, bottom line, the basalt sites are more fertile, but the rhyolite sites store more water.

Freeze/thaw cycles, erosion

All of the soils in the study sites are in the frigid soil temperature and xeric soil moisture regime and lie below the persistent snow zone, which makes them more susceptible to rain-on-snow events and subsequent erosion. Without an insulating cover of snow, day and night soil temperatures oscillate widely, cycling each day from below freezing temperatures at night to thawing by mid-day. As the water in the soil freezes, the expanding ice lifts the soil above it. As the ice thaws, the soil collapses downward again. When this occurs on a slope, the process is called solifluction. With each thaw cycle, the soil settles a bit downslope in response to gravity, creating a net downslope movement. The little puddle of soil on the surface is particularly vulnerable to overland water flow carrying it further downslope. Freeze/thaw cycles or the looseness of rhyolitic gravel result in thinner cryptogam crusts to protect the soil from erosion (David 2013). Diurnal freeze/thaw cycles repeated every winter for thousands of years were part of the erosion process that created the scablands.

Rocks and root stability

The roots of the various grass species found on scablands differ from each other. These differences interact with the characteristics of the rocks. Binney (1997) found that Henderson's ricegrass roots are much longer than roots of its most common associate, Sandberg bluegrass (*Poa secunda*). Although the roots of Sandberg bluegrass are shallower, we observed that they are much more densely branched. Henderson's ricegrass roots have a cork-like layer in the root cortex which allows the outer covering of the root to move when the soil expands and contracts, without damaging the inner root (Maze 1981). The root remains anchored and able to take up water and nutrients from the soil despite frost action that moves the soil. Roots of the other common grass on the scablands, one-spike oatgrass (*Danthonia unispicata*), have a branching pattern similar to Henderson's ricegrass.

On the basalt scablands, rock cover ranges from scattered to nearly total cover of the soil. Henderson's ricegrass usually grows as isolated clumps amongst a tight matrix of cobbles and stones solidly embedded in the soil surface. In contrast, Sandberg bluegrass and one-spike oatgrass do not show a strong association with the microsites created by dense rock cover. Wedged tightly together, the heavy basalt rocks are not easily lifted by ice in the soil, but just as importantly they insulate the soil underneath from temperature fluctuation, damping the freeze/thaw cycles (Poesen and Lavee 1994). In addition, as the roots of the ricegrass grow deeper into the soil, they probe for fractures in the underlying bedrock. The roots follow these cracks to access additional water and further stabilize the plant.

Because the rhyolitic tuff is lighter and stores more water it is more subject to frost-heaving than the basalt. We saw much more frost-heaving of grass bunches growing in bare soil or with a loose rhyolitic gravel mulch than in the basalt rock matrix. Sandberg bluegrass plants are commonly thrust out of the ground or left perched on pedestals by frost heave events. However, this is less serious for the bluegrass than it is for the ricegrass. Densely branching roots cling to enough soil to allow it to survive through flowering and seedset in early spring. By the time the exposed soil dries out, the bluegrass has gone dormant for the summer. In contrast, Henderson's ricegrass plants depend on deep soil moisture because they flower in June and ripen seed during the summer drought. When Henderson's ricegrass is heaved out of the ground, its roots are unable to retain the uplifted

Eriocoma hendersonii roots are less densely branched than roots of *Poa secunda*. Note: the full root length was not excavated. Photo by Robert Korfhage.

Densely branched fibrous roots of *Poa secunda*. Photo by Robert Korfhage.

The density of root branching in *Danthonia unispicata* more closely resembles roots of *Eriocoma hendersonii* than those of *Poa secunda*. Photo by Robert Korfhage.

Frost-heaved clump of *Poa secunda*. Photo by Robert Korfhage.

When the soil washes or blows away from the exposed roots, the ricegrass plant dies. Photo by Robert Korfhage.

Wild ungulates pull up plants when the soil is moist in late winter/early spring. Photo by Cindy Roché.

soil. When wind or water expose the roots going into the summer drought, the ricegrass plant dies.

Plant phenology and stored soil moisture

In addition to providing insulation that dampens temperature swings, rocks prevent surface evaporation, so that soil underneath stores more moisture (Poesen and Lavee 1994). Because Sandberg bluegrass flowers early in the spring and is dormant during the hot dry summer, it doesn't require as much stored moisture. Ricegrass plants flower later than bluegrass and depend on deep soil moisture to ripen seed during the summer drought. The rock matrix may also contribute to successful seedling establishment. As seeds falls between anchored rocks, they produce seedlings that are protected from frost heaving, soil erosion, and surface evaporation. The rocks limit the number of adjacent competitive plants, a clear benefit to the ricegrass. Using common garden experiments, Binney (1997) found that Henderson's ricegrass had a slower growth rate than Sandberg bluegrass and that the ricegrass appeared to grow best without any neighboring plants, regardless of whether the neighbors were ricegrass or bluegrass.

Rocks and accessibility

The rhyolitic sites are more accessible to ungulates and off-road vehicles, providing a more comfortable surface for hooves and tires than angular basalt rocks, especially sites with boulders (rocks > 400 mm, 16 inches). Jim David observed that elk, mule deer, and antelope graze Henderson's ricegrass as soon as snow melts from the rhyolitic scablands. When the soil is moist, grazing ungulates can easily pull plants out of the loose gravelly soil and leave them exposed on the soil surface. In contrast, he seldom saw ricegrass plants pulled up in the basalt scablands. Roots confined to interspaces between large rocks require more force for removal than from a loose gravelly mulch. In addition, the matrix of angular rocks with little open soil protects plants from trampling and grazing by domestic livestock, particularly cattle. Rhyolitic sites are comparable to bare soil scablands in their vulnerability to livestock trampling and grazing and off-road vehicles.

Nature's energetic excavators: pocket gophers

Northern pocket gophers (*Thomomys talpoides*) are a common inhabitant of scablands in eastern Oregon. These fossorial (digging) rodents profoundly affect the microtopography, soils, plants, and other animals in their environments. Their tunnels help alleviate soil compaction by increasing aeration and water infiltration and producing fluffy mounds of soil on the surface.

Pocket gophers are active year-round, excavating systems of tunnels that they maintain to obtain and cache food. In the winter they may tunnel through the snow to obtain additional food plants; snow tunnels are later filled with soil from the underground tunnels. Digging and tunnel maintenance burns massive amounts of energy: estimates of the energy cost of burrowing range from 360 to 3400 times that of aboveground travel (Huntly and Inouye 1988). Pocket gophers lift between 1 and 8.5 kg/m²/year of soil to the surface (Huntly and Inouye 1988). As a result, energy flow through pocket gopher

Pocket gopher mound on a rhyolitic site. Photo by Robert Korfhage.

After the snow melts, soil-filled gopher tunnels appear on the soil surface. Photo by Cindy Roché.

populations is comparable to that of some large grazers. Not surprisingly, gophers prefer tap-rooted forbs over fibrous-rooted grasses, especially plants with succulent / starchy belowground storage organs like *Lomatium* and *Lupinus* species (Cox 1989). Over long periods of time, as gophers remove soil from under rocky areas and deposit it on adjacent terrain, they sort the rocks from the fine soil particles. As the rocks become more concentrated without underlying soil, they collapse to form swales, which channels runoff that increases the removal of soil. Their soil mining activities are credited for creating sharply defined beds of sorted stones that encircle the mounds of biscuit and swale scablands in eastern Oregon (Cox and Allen 1987). In Wasco County, Oregon, Henderson's ricegrass is found only in these stony swales and not on the adjacent mounds. Populations of fleshy-rooted plants (*Allium*, *Lomatium*, *Lupinus*, *Trifolium*) also increase where the stony soil protects the roots from gopher herbivory. Compared to the finely fibrous roots of Sandberg bluegrass, the corky roots of Henderson's ricegrass may appear positively succulent to pocket gophers. Although less preferred than succulent forbs, ricegrass plants establishing in soils with lower rock content would be vulnerable to gopher herbivory. Annual and short-lived plants thrive with gopher disturbance of the soil, especially members of the sunflower family. In contrast, long-lived slow-growing perennials (*Eriocoma*, *Lomatium*, and *Allium*) decrease with gopher herbivory. Fewer of those plants grow on the rhyolitic tuff sites and less rocky scablands, which are attractive sites for gophers.

Henderson's ricegrass in a changed world

We observed high mortality of Henderson's ricegrass on the rhyolitic tuff sites and dwindling populations on scablands without dense rock cover. Our conclusion is consistent with a 2013 report assessing the status of Henderson's ricegrass (Farris 2013). Of the 49 populations known in Oregon, 48 occur on federal land, with 33 sites on the Ochoco National Forest and 15 sites on the BLM Prineville District. In 2006, 17 populations on the Ochoco National Forest were revisited; populations on 15 of those sites had decreased since the previous census in the early 1990s (Farris 2013). In a little over 10 years, five of the fifteen sampled populations had plummeted from over 150 individuals to zero. The rapid downward trend suggests that the species was more abundant and widespread prior to the introduction of livestock and invasive annual grasses on western rangelands. While a changing climate with more severe and frequent droughts may be one factor causing the decline of Henderson's ricegrass, domestic livestock grazing, soil erosion and loss of the cryptogam layer, noxious weed competition, and human recreational activities are also major contributors.

Sheep grazing in the Ochocos, 1900. Photo courtesy of the Bowman Museum, Prineville.

Livestock grazing: the cryptogam crust, soil erosion, and invasive annual grasses

Dr. Fred Hall, with dual doctorates in Plant Ecology and Range Management, established the first ecology program in the Forest Service, in the Pacific Northwest Region. As a prominent range ecologist in eastern Oregon, he gathered extensive data for over 50 years. During the years before and after Dr. Hall's retirement, Jim David assisted him in organizing several decades of data from long-term research plots throughout the Ochoco National Forest. Dr. Hall believed that the impacts of heavy grazing in the Ochocos were never fully appreciated by the younger generations of resource managers. Populations of some rare species were permanently eliminated or reduced to a few remnants during this period of overgrazing. Examples of extirpated grasses in Oregon and Washington include *Agrostis hendersonii*, *Eragrostis lutescens*, and *Sporobolus neglectus*. *Melica smithii* and *Muhlenbergia minutissima* have been reduced to only one or two known populations in Oregon. In addition to direct damage to rare native species by grazing, livestock also contributed to the loss of the cryptogam layer, soil erosion, and the introduction of aggressive exotic plants adapted to disturbance.

History of livestock introduction

Starting in the mid- to late 1800s, the Ochoco Mountains were severely grazed by domestic livestock, primarily sheep and cattle, but also horses. Large bands of sheep

Moss may dominate the cryptogam layer; here the dark surface is moss and the rhyolitic rocks appear light. The other common plant is bighead clover (*Trifolium macrocephalum*). Photo by Cindy Roché.

stripped the rangelands bare and were often bedded on the flatter ground (scablands) along the spine of the Ochocos. By 1900, even the cattlemen were adding sheep to their eastern Oregon operations, believing that "cattle and sheep together used range better than either alone; they ate different things …What they did not realize was that together the two stripped the range clean, speeding the environmental degradation that would soon plague cattle- and sheepmen alike" (Cox 2019, p. 63). Shaniko, the county seat of Wasco County was "the Wool Capital of the World." In 1903 alone, nearly 5 million pounds of wool were shipped by rail out of the town (Cox 2019).

Cryptogam layer and soil erosion

The cryptogam crust, a fragile layer over the surface of the soil, is made up of mosses, lichens, algae and bacteria. The crust is important for retaining soil moisture by impeding evaporation. It contributes to soil fertility and provides microsites for establishment of native plant seedlings. In her study of scabland habitat of Henderson's ricegrass, Binney (1997) found greater diversity in the vegetation on scablands with an intact cryptogamic crust than where the crust was disturbed. Continued livestock grazing and the impact of recreational vehicles prevent recovery of cryptogam crusts destroyed by past excessive grazing.

The cryptogam crust also protects the soil surface from erosion by wind and water. Fred Hall told Jim David that a particular black lichen (shown in the photo) requires at least 50 years to cover the bare surface of a rock. The area of bare rock between the lichen and the soil surface indicates either erosion or frost heaving. It is like a high-water mark on rocks along a river or lake. The bare space below the lichen may indicate one to two inches of soil lost at the rhyolitic tuff site where the rock was photographed.

Invasive annual grasses

Invasive annual grasses, particularly ventenata (*Ventenata dubia*), have been implicated in the decline of Henderson's ricegrass on the Ochoco National Forest (Dewey 2013). Hall noted that the annual bromes and ventenata were introduced to the Ochocos during the period of habitat degradation a 100 to 150 years ago. In 50 years of observing eastern Oregon rangeland, Hall saw weedy annual grass populations gradually decline during decades of improved management of the cattle and sheep. He also watched annual plant populations expand and contract with changes in precipitation (drought or above normal), soil disturbance, increased or decreased competition from perennial native species, and fire cycles. On seasonally saturated (clay influenced) soils, ventenata and medusahead (*Taeniatherum caput-medusae*) are more competitive than annual bromes. Two sources indicate the presence of ventenata since the 1950s and 60s. When Hall worked on the Ochoco National Forest in the mid to late 1950s, he regularly flew his plane to the old airstrip at Cinnabar Flat on the Paulina Ranger District. At that time, he recorded the presence of ventenata on the scabland around the airstrip. He believed that at some locations ventenata was probably misidentified as *Deschampsia danthonioides*, a native annual of seasonally saturated soils. His observation is supported by aerial photos from the 1960s in which Jim David saw areas of dense fine-textured light-colored annual grass with the appearance of ventenata. Both ventenata and medusahead have been increasing on scabland sites in recent decades (Dewey 2013).

The lower edge of the black lichen indicates the level of the soil surface (prior to erosion); the pale area was previously covered by soil. Photo by Robert Korfhage.

Why are we sharing our observations?

Henderson's ricegrass appears to be retreating to extremely rocky refuges in the Ochoco scablands. Time may be running out for Henderson's ricegrass in Oregon, so we hope to inspire research that focuses on specific habitat requirements of both Henderson's and Wallowa ricegrass. Investigations might test our interpretations of site characteristics (rocks, frost heaving, underground herbivory) in relation to the population dynamics of the species, leading to recommendations for better management of the habitat and conservation of the species.

In support of our suggestion that scientists choose the ricegrasses for a research subject, Jack Maze offers some time-tested advice that worked well for him: choose a plant that grows in a place that's beautiful. These two ricegrasses meet that criterion exceptionally well. He reminiscences about sitting in the open area below Skookum Rock or on the ridge above Boner Gulch, just looking and enjoying. He captured the experience in this Haiku:

And to distant mountains
forests and grassland—
comfort and freedom.

Wallowa ricegrass (*Eriocoma wallowaensis*). Photo by Robert Korfhage.

Overview of scablands along the North Fork of the Crooked River, Ochoco National Forest. Photo by Robert Korfhage.

The Naming of Henderson's Ricegrass and Wallowa Ricegrass

Oryzopsis hendersóni was named by George Vasey (1822-1893), curator of the US National Herbarium, for Louis Henderson, who collected it on dry, rocky ground 14 June 1892 near the summit of Mount Clements [Cleman Mountain, northwest of Naches, Washington] (Vasey 1893, Hitchcock 1971). Marcus E. Jones (1912) reduced it to a variety of *O. exigua*, but Mary Barkworth (1993) reinstated it as a species with the name *Achnatherum hendersonii* 100 years after Vasey first described it. It now bears the name *Eriocoma hendersonii*, having been moved to the "new" genus along with the other species previously called *Achnatherum* (Peterson *et al.* 2019).

In 1960-61 Jack Maze was a graduate student in the Department of Botany at the University of Washington, where he started to work on grasses that were then called *Stipa*. He did several different types of studies over the next 35 years or so, but the ones that are especially memorable involve Oregon. It started with Henderson's ricegrass, an endemic with a restricted distribution on scablands in central Oregon and the east slope of the Cascades in Washington. From 1990-93 he and Kathleen (Kali) Robson were doing research with the short-lived Rare Plant Consortium at the Forestry Sciences Lab in Wenatchee, Washington. One plant of interest was Henderson's ricegrass, which was on the "Watch List" because of its restricted distribution.

In addition to the sites in Washington, much of Jack's and Kali's research was done in two National Forests in Oregon: the Ochoco and Wallowa-Whitman. On one collecting trip, Wallowa-Whitman Forest north zone botanist Marty Stein took them to some of their Henderson's ricegrass sites. They noticed the plants were a bit different from the Henderson's ricegrass plants in other places. Their inflorescences were laxer and more open, the panicle branches drooping. It looked to all of them like it might be a new species, so they asked Marty if he wanted to describe it and offered to

help. He declined, so Jack and Kali published it as a new species, *Achnatherum wallowaensis* (Maze and Robson 1996). They collected the type specimen near Boner Gulch on the Wallowa-Whitman National Forest, and paratypes from other locations on that Forest and from the Ochoco National Forest. The name has since been changed to *Eriocoma wallowaensis* (Maze and Robson) Romansch (Peterson *et al.* 2019).

Type specimen of *Eriocoma hendersonii* collected by L.F. Henderson. Image courtesy of the US National Herbarium, Smithsonian Institution.

Acknowledgments

In her "Adopt-a-Scab" project, Jill Welborn, former botanist on the Paulina Ranger District, introduced the scablands to Cindy, leading to a fascination with the elusive ricegrasses and ultimately this article. Bob Korfhage gamely participated in many camping adventures to the Ochoco Mountains and photographed plants and habitats. Sue Vrilakas (ORBIC) and Sarah Canby (Prineville District BLM) assisted with locations. Elizabeth Binney, Paula Brooks, Jill Welborn, and Sue Vrilakas reviewed the manuscript. Cindy credits her long-time friend Tom Brannon, who succinctly summed up Henderson's ricegrass habitat on the Colockum (northcentral Washington) with just four words: "more rocks than soil." Kareen Sturgeon provided editorial advice on numerous draft versions. We dedicate the article to Marty Stein (1955-2022), who died on May 3.

References

Barkworth ME. 1993. *Achnatherum hendersonii* (Vasey) Barkworth. Phytologia 74(1):7.

Barkworth ME, Capels KM, Long S, Anderton LK, Piep MB. 2007. Stipeae. *Flora of North America Vol. 24.* Oxford University Press, New York. pp. 109-186.

Binney EP. 1997. Comparative analysis of community and population levels of organization in the rare grass *Achnatherum hendersonii*. PhD Thesis, University of British Columbia, Vancouver, Canada.

Cox GW. 1989. Early summer diet and food preferences of northern pocket gophers in north central Oregon. Northwest Science 63(3):77-82.

Cox GW, Allen DW. 1987. Sorted stone nets and circles of the Columbia Plateau: a hypothesis. Northwest Science 61:179-185.

Cox TR. 2019. *The Other Oregon. People, Environment, and History East of the Cascades.* OSU Press, Corvallis. 398 pp.

David J. 2013. Henderson's ricegrass (*Achnatherum hendersonii*) – Soil & Site Relationships. Unpublished report, Ochoco National Forest, Prineville, Oregon.

Dewey R. 2013. Conservation Assessment for Henderson's Needlegrass (*Achnatherum hendersonii*) and Wallowa Needlegrass (*Achnatherum wallowaense*). USDA Forest Service Region 6 and USDI BLM, Oregon and Washington. (https://www.fs.fed.us/r6/sfpnw/issssp/documents2/ca-va-achnatherum-species-2013-04.docx)

Farris K. 2013. Long Term Monitoring and Habitat Assessment of Henderson's and Wallowa Needlegrass Populations, Final ISSSSP Report ACHE10, Region 6 Forest Service, Pacific Northwest Region.

Fryxell R. 1965. Mazama and Glacier Peak volcanic ash layers: relative ages. Science 147(3663):1288-1290.

Hitchcock AS. 1971. *Manual of the Grasses of the United States.* 2nd Ed. (rev. by Agnes Chase). Dover Publications, New York. Vol. 1, p. 438.

Huntly N, Inouye R. 1988. Pocket gophers in ecosystems: patterns and mechanisms. BioScience 38(11):786-793.

Johnson CG, Clausnitzer RR. 1992. Plant Associations of the Blue and Ochoco Mountains. USDA Forest Service Pub. R6-ERW-TP-036-92. p. 2.

Jones ME. 1912. *Oryzopsis exigua* var. *Hendersoni.* Contrib. West. Bot. 14:11.

Maze J. 1981. A preliminary study on the root of *Oryzopsis hendersonii* (Gramineae). Syesis 14: 151-154.

Maze J, Robson KA. 1996. A new species of *Achnatherum* (*Oryzopsis*) from Oregon. Madroño 43:393-403.

Nuttall T. 1818. *Eriocoma cuspidata* Nutt. Genera of North American Plants. 1:40-41.

Paulson DJ. 1977. Ochoco National Forest Soil Resource Inventory. USDA Forest Service, Pacific Northwest Region. 291 pp. plus 56 mapping sheets.

Peterson PM, Romaschenko K, Soreng RJ, Reyna JV. 2019. A key to the North American genera of Stipeae (Poaceae, Pooideae) with descriptions and taxonomic names for species of *Eriocoma, Neotrinia, Oloptum,* and five new genera: *Barkworthia,* ×*Eriosella, Pseudoeriocoma, Ptilagrostiella,* and *Thorneochloa.* PhytoKeys 126:89-125.

Poesen J, Lavee H. 1994. Rock fragments in top soils: significance and processes. Catena 23:1-28.

Rydberg PA. 1912. *Eriocoma hymenoides* (R. & S.) Rydb. comb. nov. Bull. Torrey Bot. Club 39:102.

Vasey G. 1893. *Oryzopsis hendersoni* Vasey. Contr. U.S. Natl. Herb. 1:267

Vrilakas S. 1990. Draft species management guide for *Oryzopsis hendersonii.* Part of the challenge cost share program with Wallowa-Whitman NF, Oregon Dept. of Agriculture, and Oregon Natural Heritage Program. Document on file at ONHP and ODA.

Winward A, Youtie B. 1976. Ecological inventory of the Lawrence Memorial Grassland Preserve. Unpublished document on file at the Nature Conservancy, Portland, Oregon.

Cindy Roché joined the Native Plant Society of Oregon in 1998 when she moved to Oregon from Washington state, where she earned BS and MS degrees from Washington State University and a PhD at the University of Idaho. Her interest in grasses dates back to the late 1970s working as a range conservationist with the Forest Service. She illustrated grasses for volumes 24 and 25 of the *Flora of North America*, published in 2003 and 2007. More recently, she collaborated with the Carex Working Group and Robert Korfhage to publish a *Field Guide to Grasses of Oregon and Washington*, published in 2019 by OSU Press. She started teaching grass identification at the Siskiyou Field Institute in 2005 and continues to offer intensive field workshops on grass identification. She and her husband enjoy exploring Oregon and Washington in search of rare grasses and other adventures.

Jim David was born in John Day, Oregon, and raised in the Klamath Mountains of California. He earned a BS in Range and Wildland Science and a MS in Range Ecology at the University of California, Davis. He has worked as a soil scientist with the BLM in Nevada with specific experience in soil/vegetation mapping, hydrology and range condition mapping which included evaluating the effects of cattle, sheep, and wild horses on rangelands. For over 30 years he has been working on the Ochoco National Forest and Crooked River National Grassland as the Forest Soil Scientist, which has included mapping soils and vegetation relationships for use and management interpretations.

Jack Maze was born in San Jose and raised in Hollister, California, birthplace of the American biker. He has a BA in Biology from Humboldt State University (College when he was there), an MS in Botany from the University of Washington with CL Hitchcock as his thesis supervisor, and a PhD in Botany from the University of California, Davis. He has done research in taxonomic revision in grasses, distributional changes during the Pleistocene, population differentiation in *Abies,* plant development, plant morphology, the expression of increasing complexity that accompanies development and evolution and the relationship between ontogeny and phylogeny. He is a Professor Emeritus of Botany at the University of British Columbia and resides in Vancouver, British Columbia.

Tanya Harvey

Tanya grew up in New England, the daughter of avid gardeners. She's been in love with nature, especially plants, as long as she can remember. As a child, she spent as much time out of doors as possible, driven to learn all she could about the natural world. She particularly recalls going on a nature hike and wanting to be like the naturalist leading it. She revealed artistic talent at an early age, filling notebook after notebook with sketches and watercolors. Although she obtained a BA in Mathematics from Dartmouth College in 1980, it was her early passions that led to her career as a multimedia artist and designer who is inspired by nature and the outdoors.

She has held a succession of paid and volunteer positions in graphic design and publishing. She sells and displays her art, craft, and design work on tanyaharveydesign. com, and for many years sold her work at the annual Portland Audubon Wild Arts Festival. Lately, however, she's too busy for the festival. Since 2012 she's been employed by OregonFlora to work on the three-volume *Flora of Oregon*. She was responsible for design, layout, and editing of the two volumes already produced and is now working on the third. She also contributes photos and some illustrations. Many of her photos and species lists appear on the OregonFlora website (ore-gonflora.org). Tanya's knowledge of the flora of Oregon in general and the Western Cascades in particular is among the most comprehensive of any botanist in the state. She edited every treatment thus far submitted for inclusion in the *Flora of Oregon*, evaluating it from two standpoints: that of an end user of the *Flora* and that of a reviewer or contributor. She also does editing and layout for the remaining volumes of the *Flora of North America*.

In 1987 she met her husband, Jim Babson, in California. They moved to Oregon in 1992 and settled on 55 spectacular acres in Fall Creek, where Jim renovated a fixer-upper house. Much of the land is in a semi-natural condition, and Tanya is lovingly restoring its native vegetation, with special attention to the oak and grassland habitats. She also maintains a fenced garden, which she has filled with woodland and rock garden species, many of them native. She worked with the Middle Fork Willamette Watershed Council (MFWWC) to obtain funding for her restoration, and she recently hosted a MFWWC tour for her neighbors.

Tanya joined the Native Plant Society of Oregon (NPSO) in 1999 (soon after arriving in the southern Willamette Valley) and became a life member in 2004. As a member of the Emerald Chapter, her contributions include the following:

- Produced the monthly *NPSO Bulletin* (including both editing and layout) for nearly nine years (from April 2000 through 2008), producing almost 100 print issues;
- Designed posters, t-shirts, and more for the state board and her chapter, including for the Mount Pisgah Arboretum Wildflower Festival from 1998 to 2005;
- Chaired the field trip committee and served as the Friday night speaker for the 2008 annual meeting hosted by Emerald Chapter;
- Presented 16 slide talks to various NPSO chapters and another 17 talks on plants to other organizations in Oregon;
- Led over 20 field trips for numerous annual meetings and various NPSO chapters and botany hikes for a number of other organizations;
- Conducted a rare plant survey for Citizen's Rare Plant Watch.

Soon after joining NPSO, Tanya became active in a number of other local botanical and conservation organizations. She joined the local chapter of the North America Butterfly Association, edited the newsletter for the Eugene Hardy Plant Group and served as President of the Emerald Chapter of the North American Rock Garden Society.

Throughout all that, Tanya has taken every opportunity to botanize and photograph the Western Cascades, the oldest part of the Cascade Range. In fact, she so loves these mountains that she and Jim were married atop one! Having taken over a thousand hikes in the Western Cascades solo or with fellow plant lovers, she's a supremely authoritative field trip leader. During her explorations she has found a number of uncommon species, which she collected for the OSU Herbarium. Since 2010, she has maintained a popular website, westerncascades.com, where her jaw-dropping photographs accompany more than 300 information-packed trip reports. It also includes descriptions and her personal plant lists for numerous botanically interesting locations. Eventually, she plans to develop the information from each of the 150 or so locations she has botanized into a complete field guide to the Western Cascade flora. Her book has been on hold since she started working on the *Flora of Oregon*.

Tanya's Mountain Plants of the Western Cascades website and blog is an incredibly useful and well-organized resource for all wildflower enthusiasts. Through stunning photos, captivating natural history stories, and expert botanical knowledge, she brings the flora of this region of Oregon alive for a broad audience. When I bought acreage near hers, Tanya connected me with MFWWC's Restoration Projects manager, which advanced my savanna restoration. Tanya identified multiple native and introduced species and gave me seeds for native plants. It is a privilege to nominate Tanya for this honor she richly deserves.—*Karl Anderson, Emerald Chapter.*

The Sierra Nevada blue (*Agriades podarce*) is an uncommon butterfly of wetlands in the southern Cascades. Their caterpillars feed exclusively on species of shooting star (*Dodecatheon jeffreyi* and *D. alpinum*). Here a female is nectaring on great camas (*Camassia leichtlinii*) in the Calapooya Mountains. Photo by Tanya Harvey.

BOOK REVIEWS

A Place for Inquiry, A Place for Wonder: The Andrews Forest
William G. Robbins
2020. ISBN 9780870710193
Oregon State University Press, Corvallis, Oregon.
242 pp. 22 b&w photos, map, chart, table, index. 6 x 9 in. paper. $29.95.

William Robbins has produced the first comprehensive historical account of the origins, development, and importance of the H.J. Andrews Experimental Forest. This renowned research forest, commonly referred to as the Andrews, is located near Blue River and managed cooperatively by the Forest Service's Pacific Northwest Research Station, Oregon State University, and the Willamette National Forest. The author is an Emeritus Distinguished Professor of History at Oregon State University who has previously written about the forest. He waded through countless historical documents to glean the turning points of a 70+ year chronology. In two hundred pages, he outlines the external forces that shaped the early direction of research, highlights the most pivotal personalities, programs, and events, and traces how research emanating from this Place for Inquiry has shaped national policy.

But the wonder is missing. I was fortunate enough to experience that wonder several times. I found it amid the peace and sanctity of the old-growth forest itself while undertaking an undergraduate research project in 1991. Later, as a graduate student, I experienced it while delving into the collective wisdom of the generations of researchers who came before me. I was awed by the continuity of time and the complexity of the products of natural selection and humbled by the scale of my own contributions and the insignificance of our species in an ecosystem that has vastly less need for us than we have for it.

In the introduction, Robbins provides a dispassionate, albeit concise, summary of the history and importance of the Andrews, and the subsequent seven chapters continue to trace that history in dry detail. In the first chapter, he describes the historical context of its 1948 establishment and initiation into the International Biological Program. The subsequent multi-disciplinary ecosystem research that led to its participation in the UNESCO Man and the Biosphere and the NSF-funded Long Term Ecological Research programs is the topic of the second chapter. The origins of old-growth research on the Andrews are recorded in the third chapter, including how this research led to the listing of the northern spotted owl as an endangered species and to the advent of a new, less commodity-oriented philosophy toward forest management known as "new forestry." The fourth chapter highlights the political fallout of these developments, which culminated in a Northwest Forest Plan meant to balance needs for both wildlife habitat and economic stability. Robbins frankly reports on the success of the plan to conserve old-growth, as well as its failure to sustain timber-dependent communities.

In the fifth chapter, Robbins characterizes the adverse political climate in which research into ecosystem science took place at the end of the 20th century. He describes public and educational outreach (e.g., the NSF-funded Research Experiences for Undergraduates program in which I participated) as well as ongoing research into biodiversity. In the final chapter, Robbins documents the experiences, and quotes the resulting works, of various writers who found inspiration while in residence at the Andrews (and Mt. St. Helens) through programs promoting the intersection of the sciences and the humanities. In conclusion, Robbins notes that the decades-long research at the Andrews continues to shape forestry, scientific thought, and national and global environmental policy at a critical time for understanding the ecosystem-wide effects of climate change.

The consummate academic historian, Robbins has produced a mechanistic accounting of the times, places, events, and some of the people comprising the Andrews— which makes for a satisfactory documentary record, but a disappointing casual read. Although the elements of history have been strung together in a chronological sequence, in the absence of a narrative thread readers are

left to piece together linkages on their own. The extensive but dusty documentation from which this history was extracted is presented with unnecessary repetition and factual quotations that provide little insight. Historical tidbits suddenly appear without preamble, context, or interpretation, choppily tied together. Periodic attempts to humanize the text by interspersing personal accounts seem to be detached digressions, only tangentially related

to the Andrews. Even the quotations of artists who spent time at the Andrews fail to convey the sense of wonder their visits must surely have inspired.

Which is a little surprising, because wonder flows from at least two wellsprings on the Andrews. I can appreciate how a casual visitor might miss the academic wellspring—perhaps only a fellow researcher can feel humbled and awed by the accumulated wisdom and voluminous research spanning over seven decades. But the wonder and majesty of the old-growth forest underlying this research is easily accessible to the public through the Lookout Creek Old-Growth trail. The trail winds through the forest under towering 500-yr-old trees too large to hug and alongside massive logs rotting so slowly that they remain obstacles for centuries after toppling to the ground. To be fair, academicians must walk a tightrope between objectivity and passion, but the products of the former should not be mistaken for popular literature. Read *A Place for Inquiry, A Place for Wonder* to understand how it came to be a place for inquiry but *visit* the Andrews to discover it as a place for wonder.—*Jeri Peck, former Oregonian, now Research Associate at Pennsylvania State University*

The View from Cascade Head: Lessons for the Biosphere from the Oregon Coast
Bruce A. Byers
2020. ISBN 9780870710353
Oregon State University Press, Corvallis, Oregon.
216 pp. 16 B/W illus. 1 map. 6 x 9 in. paper. $22.95.

The frontispiece reveals that the name *Cascade Head* in the book title refers to the Cascade Head Biosphere Reserve, which encompasses a number of discrete components, including the Neskowin Crest Natural Research Area, Cascade Head Experimental Forest, The Nature Conservancy Cascade Head Preserve, Sitka Center for Art & Ecology, Salmon River and Estuary, Camp Westwind, and Cascade Head Marine Reserve.

In 15 chapters, Byers tells the story of the Biosphere Reserve, arranging his narrative by he calls the "re" words:

resistance, research, restoration, reconciliation, and resilience. Byers uses the metaphor of "the eagle's view" to unite his essays around these themes and to explore ethical implications for our human behavior and that of our culture. The theme of resistance speaks loudly and clearly as we learn how, in 1973, local individuals (including Senator Bob Packwood), alarmed at encroaching development surrounding Cascade Head, a basalt headland with dramatic ocean views, laid the foundation for what would become the reserve. Their efforts illustrate the first of three lessons presented in the book: the commitment and hard work of individuals can make all the difference.

The Biosphere Reserve encompasses 160 square miles, extending along the central Oregon coast between Neskowin and Lincoln City. In beginning chapters, Byers traces the history of the biosphere concept (it dates to 1890 in Ukraine), a worldwide program "to create a network of places dedicated to monitoring and understanding the diverse ecosystems of the biosphere and developing models and strategies for maintaining or restoring their resilience while still meeting human social, cultural, and economic needs." Having worked in 34 biosphere reserves in 17 countries, Byers is eminently qualified to reveal the lessons that Cascade Head has to offer.

Over 40 years ago, Byers and I shared an advisor in our graduate programs at the University of Colorado in Boulder. Later, as a biology professor at Linfield College in McMinnville, I led field trips for my students to The Nature Conservancy's Cascade Head Preserve to study the ecology of temperate rainforests. My plant taxonomy students and I spent weekends at Camp Westwind learning to recognize coastal species. Over the years, I have taken several courses at the Sitka Center for Art & Ecology where Byers was an Ecology Resident in 2018, and we both share a fondness for the molasses bread at the Otis Café. I know this area well, but until I read this book, I didn't realize how much more there was to learn.

Byers introduces us to a host of ecological concepts, such as succession, species and genetic diversity, nutrient cycling, predator-prey relationships, decomposition, evolution, and symbiosis. We learn the ecological importance of keystone species, such as sea stars that, as dominant predators, maintain species diversity in intertidal ecosystems. Byers documents extensive research that revealed the importance of protecting mature forests for the survival of an entire group of species, including the marbled murrelet, spotted owl, and red tree vole. He outlines how restoration of the Salmon River Estuary was critical to the survival of juvenile salmon and returning adults. Yet, despite extensive research, he highlights a second lesson, that "ecological mysteries" still abound. What explains migratory patterns of grey whales? What caused sea star wasting, and why did the disease suddenly disappear? What is the best way to restore genetically distinct populations of Oregon silverspot butterflies to increase their resilience in coastal prairies?

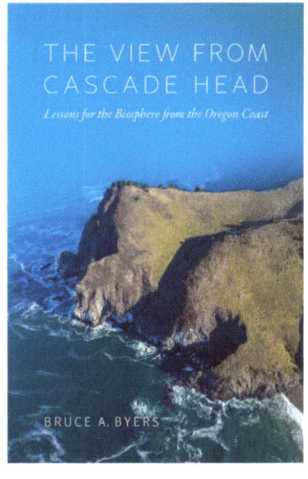

Byers demonstrates that, over time, researchers have had to increase the scale of their studies from local sites to a landscape level of analysis. For example, research on beavers revealed their critical role in maintaining forest health by regulating the hydrologic cycle on land, reducing runoff and sedimentation of streams. In turn, healthy streams facilitated the return of salmon from the ocean to their natal streams where, upon death, they returned nutrients to the forest. Recognizing this connection between the land and the sea led ultimately to the addition of the Cascade Head Marine Reserve to the larger Biosphere Reserve.

Byers describes a "multi-use shades of green landscape model," a worldview where ecological, economic and cultural benefits are balanced among stakeholders. He concludes that how we think about our human relationship to nature shapes our individual and collective actions. This third lesson to be gained from the book offers a pathway for reconciliation between often competing interests and opposing worldviews. He shares how his residency at the Sitka Center for Art & Ecology revealed how art and ecology intersect, that art is "one aspect of human ecology, and ecological science is a kind of art." In this interpretation, art is, like science, an adaptive behavior, helping to protect the biosphere.

The book is attractive, easy to read, and graced with lovely illustrations. A timeline table would have helped me follow the incremental addition of the various components that now make up the reserve and, of course, photos of this beautiful site would have been welcome (but would have made the book much more expensive). The only error I found was in reference to Indian hemp (*Apocynum cannabinum*), which Byers incorrectly described as a noxious weed. Its rhizomatous habit makes it troublesome in gardens and agricultural settings, but this native species is not on the Oregon Department of Agriculture's list of noxious weeds.

I recommend this book to anyone interested in gaining a deeper understanding of how the Cascade Head Biosphere Reserve has achieved the goals of UNESCO's Man and the Biosphere Program by "becoming laboratories for understanding complex social-ecological systems and models for resolving problems, restoring ecological functions and services, and increasing resilience in the face of climate change and other unpredictable events."
—*Kareen Sturgeon, Cheahmill Chapter.*